Henry Minchin Noad

The Inductorium

Induction Coil

Henry Minchin Noad

The Inductorium
Induction Coil

ISBN/EAN: 9783744678520

Printed in Europe, USA, Canada, Australia, Japan

Cover: Foto ©berggeist007 / pixelio.de

More available books at **www.hansebooks.com**

THE

INDUCTORIUM,

OR

INDUCTION COIL;

BEING

𝕬 Popular Explanation

OF THE

ELECTRICAL PRINCIPLES

ON WHICH IT IS CONSTRUCTED.

WITH THE DESCRIPTION OF A SERIES OF BEAUTIFUL AND
INSTRUCTIVE EXPERIMENTS

ILLUSTRATIVE OF

THE PHENOMENA OF THE INDUCED CURRENT.

By HENRY M. NOAD, Ph. D., F.R.S., F.C.S., &c.,

LECTURER ON CHEMISTRY AT ST. GEORGE'S HOSPITAL.
Author of "A Manual of Electricity," &c., &c.

SECOND EDITION.

PUBLISHED FOR THE PROPRIETOR BY

JOHN CHURCHILL & SONS, NEW BURLINGTON STREET,

M.DCCC.LXVI.

PREFACE TO SECOND EDITION.

THE object of this little book is to place in the hands of persons possessing or desiring to possess an Induction Coil (or *Inductorium*, as it has been called by the German Physicists) a clear and succinct account of the Electrical principles on which the instrument is constructed. Also to describe the various apparatus used, and the principal experiments to be made therewith.

It has been written at the request of, and for Mr. Ladd, the well known successful manufacturer of these machines. That some such work was wanted has been fully proved by the rapid sale of the first impression. In the present Edition, without materially increasing the size of the book, I have endeavoured to trace the progress of the development of this wonderful instrument, which promises to become a powerful means of research in many physical inquiries.

INDUCTORIUM, OR INDUCTION COIL.

1.—DISCOVERY OF ELECTRO-MAGNETISM.

In the year 1820, Professor Oersted, of Copenhagen, announced his famous discovery of the reciprocal force exerted between magnetic bars and wires uniting the opposite terminals of a voltaic battery, and thus laid the foundation of a new science—that of Electro-Magnetism. The discovery of the Danish philosopher was thus simply stated:—When a properly-balanced magnetic needle is placed in its natural position in the magnetic meridian, immediately under, and parallel to, a wire along which a current of voltaic electricity is passing, that end of the needle which is situated next to the negative side of the battery immediately moves to the *west;* if the needle is placed parallel to and over the wire, the same pole moves to the *east.* When the uniting wire is situated in the same horizontal plane as that in which the needle moves, no declination takes place, but the needle is inclined, so that the pole next to the negative end of the wire is depressed when the wire is situated on the west side, and elevated when situated on the east side. To assist the memory in retaining the directions of these deviations, Ampere devised the following formula:—" Let any one identify himself with the current, or let him suppose himself lying in the direction of the positive current, his head representing the copper and his feet the zinc plate, and looking at the needle; its *north* pole will always move towards his right hand."

2.—Electro-Magnetic Rotation.

Reasoning on the fact that this action of a conducting wire on a magnet is not a directly attractive or a repulsive one, Faraday was led to the conclusion that if the action of the voltaic current could be confined to one pole of the magnet, that pole ought, under proper conditions, to rotate round the wire; and conversely, if the magnet were fixed and the conducting wire moveable, the wire ought to rotate round the magnetic pole; both of these phenomena he realised, and described the apparatus for exhibiting them in the "Quarterly Journal of Science," Vol. xii., p. 283 (January, 1822). Ampere subsequently caused a magnet to rotate round its own axis; and Barlow devised an ingenious apparatus for exhibiting the rotation of a conducting body round its axis.

3.—The Galvanometer.

Shortly after the discovery of Oersted, Schweigger, a German physicist, applied it to the construction of an apparatus for indicating the direction and measuring the intensity of voltaic currents. This instrument is called the multiplier or rheometer, or more popularly the galvanometer. In its original form it consisted of a rectangular coil of silk or cotton-covered copper wire, in the centre of which was suspended, on a pivot, a magnetic needle, and a card graduated into 360 degs.; the instrument being so placed that the needle lies parallel to the coil; on causing a current of electricity to circulate through the latter, the needle becomes violently affected, even by very feeble currents, it being obvious, from a consideration of Oersted's fundamental law, that the needle, being placed between the two horizontal branches of the conducting wire, will be impelled in the same direction by the current traversing the wire above and below it. A great improvement was subsequently made in the instrument by Cumming and Nobili, who applied the astatic needle to the multiplier, thereby greatly increasing its sensibility, by annulling the directive action of the earth on the needle. There appears to be scarcely any limit to the sensibility which the galvanometer may be made to attain; as far as experiment has yet gone, it increases in delicacy in proportion to the length, purity, and insulation of the copper wire composing the coil. Du Bois-Reymond constructed, for his researches on the currents of electricity existing in animal structures, a multiplier, the length of which was 16,752 feet long, and passed round the frame 24,160 times; the sensibility of this instrument is almost incredible. The galvanometer is an indispensable instrument to those engaged in electrical researches.

4.—ELECTRO-DYNAMICS.

When two wires are traversed simultaneously by an electrical current, attractions or repulsions ensue, similar to those which take place between the poles of two magnets. If the currents are moving in the same direction in the two wires, they mutually attract; if in a contrary direction, they mutually repel. This discovery we owe to Ampere, and the discussion of the phenomena to which it gave rise constitutes the science of electro-dynamics. The analogy between wires conducting electricity and magnets is strikingly illustrated by turning the wires corkscrew fashion, making them helices. A helix has, indeed, all the properties of a magnet, but the nature of the pole at either end will depend on the direction of the turns of the helix; a helix in which the turns are from left to right *upwards* is a *dextrorsal* helix; a helix in which the turns are from left to right *downwards* is a *sinistrorsal* helix. In the former, the S pole is formed at the end at which the current enters; in the latter, the N pole is formed at that extremity. The analogy extends to fracture. If a magnetic bar be broken in two, each piece is a perfect magnet, and the fractured parts have opposite poles; so it is with a helix, which, if divided in the middle, exhibits attraction between the fractured ends. If a helix be suspended vertically and loosely, its upper end being held by a binding screw, and its lower end dipping into mercury; and if a voltaic current be passed along it whilst thus suspended, there will be mutual attraction manifested between the coils, and the helix will be contracted.

5.—Ampere's Theory of Magnetism.

On the analogy which exists between helices and magnets Ampere founded his theory of magnetism. According to this theory, the phenomena of magnetism depend on voltaic currents circulating round the molecules of the magnetic bodies. In their unexcited state these molecular currents move in all directions, and thus neutralize one another; but when the bar becomes a magnet, the currents move parallel to each other, and in the same direction, and the effect produced is that of a uniform current moving corkscrew fashion round the bar, which thus becomes in effect a helix, and the attractions and repulsions of the magnet are consequences of the .actions of the currents on each other. In applying this theory to the explanation of the phenomena of terrestrial magnetism, it is necessary to suppose the incessant circulation of electrical currents round the globe from east to west perpendicular to the magnetic meridian.

6.—MAGNETISM EXCITED BY ELECTRICITY.

A consideration of the influence exerted by electrical currents on magnets, leads naturally to the conclusion that the neutral condition of bodies susceptible of magnetism would be disturbed by an electrical current, and that they would become magnetic, and the fact is easily verified by plunging the wire uniting the opposite poles of a voltaic battery into iron filings, which attach themselves to the wire, and remain adhering to it as long as the current continues to circulate, but drop off the moment the circuit is interrupted; filings of copper or tin exhibit no such action. The magnetizing power of electricity is also illustrated by winding a silk or cotton-covered copper wire round a glass tube enclosing an unmagnetized steel needle and connecting the ends of the helix with the terminal plates of the voltaic battery; the needle becomes magnetized to saturation even by a momentary passage of the current through the helix; the magnetization of the needle also takes place if, instead of passing the current from a voltaic battery along the helix, a Leyden phial be discharged through it, an interesting experiment, as proving the magnetizing power of ordinary (statical) as well as of voltaic (dynamical) electricity. The sense in which the needle will be magnetized will depend on the nature of the helix; if it be a right-handed one, as—

S   Fig. 1.   N

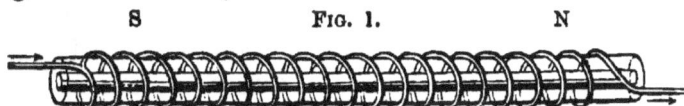

the north pole of the needle will be formed towards

the extremity at which the current enters; if the helix be left-handed, as—

N   FIG. 2.   S

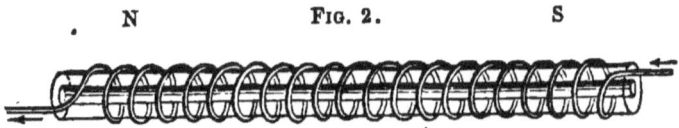

then the end of the needle nearest the extremity at which the current enters will be a south pole. A tube of wood may be substituted for one of glass in this experiment, but not one of copper, which, if thick, destroys entirely the effect of the current. The tube in this experiment may be altogether dispensed with, and the silk or cotton-covered copper wire wound round the steel bar itself, which thus becomes intensely and permanently magnetized by a very feeble current. Soft iron, treated in a similar manner, acquires a high degree of temporary magnetism—iron, if pure, not being able to retain the magnetic force, although if not pure, it does not wholly lose its polarity. Bars of iron thus temporarily magnetized by the voltaic current are called electro-magnets; they are generally made horse-shoe shape, the covered copper wire being wound several times round each arm in the same direction; the ends of the curved bar acquire opposite magnetic polarity. A convenient arrangement of the electro-magnet is

FIG. 3.

shown in Fig. 3 ; the iron cores are $10\frac{1}{2}$ × 2 inches, and are each covered with 120 yards of copper wire, $\frac{1}{10}$th of an inch in diameter. It is provided with conical soft iron armatures for convenience in diamag-netic experiments. The power of the electro-magnet depends on the dimensions and purity of the iron, the intensity of the current, and on the length and thickness of the wire. It has been shown by Dub, that the power of the electro-magnet to effect a mag-netic needle and to sustain weights, is proportional to the square root of the diameter of the bar. The applications of the electro-magnet to electro-tele-graphy, to the construction of electro-magnetic and horological machines, and to the elucidation of the phenomena of diamagnetism, have received import-ant developments during the last few years.

7.—Volta-Electric Induction.

On the 21st of November, 1831, the first series of Faraday's celebrated "Experimental Researches in Electricity" was read before the Royal Society. It related to the induction of electric currents. Two hundred and three feet of copper wire in one length were coiled round a large block of wood; other two hundred and three feet of similar wire were inter-posed as a spiral between the turns of the first coil, and metallic contact everywhere prevented by twine. One of these helices was connected with a galvano-meter, and the other with a powerful voltaic battery. When contact was made, there was a sudden effect at the galvanometer, and a similar slight effect when the contact with the battery was broken ; but whilst the voltaic current was continuing to pass through the one helix, no disturbance of the galvanometer took place, nor could anything like induction upon the other helix be perceived. The force of the induced current, which partook of the nature of an electrical wave, produced by the discharge of a common Leyden jar, was greater on breaking than on making contact; the direction of the current was, on making contact, the reverse of that of the inducing current, but on breaking contact it was in the same directions as that of the battery current.

8.—MAGNETO-ELECTRIC INDUCTION.

When the helices of copper wire were wound round a ring of soft iron, but carefully insulated from it, the effects were far greater, but they were not more permanent, the galvanometer needle speedily reassuming its natural position while battery contact was maintained, but being again powerfully deflected in the contrary direction the instant the contact was broken. When the ends of the helices were tipped with charcoal, a spark could be obtained at the moment of making contact between the other helix and the poles of the battery. Faraday next wound a similar series of helices round a hollow cylinder of pasteboard, and connected the respective ends with the galvanometer. He then introduced into the axis of the cylinder a soft iron bar, which he made temporarily magnetic, by bringing its opposite ends into contact with the opposite poles of two powerful bar magnets; the moment this was done, the needle was deflected; continuing the contact, the needle became indifferent, and resumed its first position. On breaking contact it was again deflected, but in the opposite direction, and then it again became indifferent. When the magnetic contacts were reversed, the deflections were reversed likewise. Here, then, was a distinct evolution of electricity from magnetism alone. The action of the current from the voltaic battery is called *volta-electric induction;*

that produced by the magnet is called *magneto-electric induction*, and the reason why the effects at the galvanometer are so much greater when the helices are arranged round an iron bar, than when they are coiled simply round a block of wood, is because, in the former case, we have volta-electric and magneto-electric induction combined, whereas, in the latter case, the effect is due to the action of volta-electric induction only. Powerful effects at the galvanometer were obtained on bringing the ends of the system of helices with an enclosed iron cylinder, between the poles of a strong magnet, and even when the coil, without the iron core, was introduced between the poles of the magnet, but without touching, so that the only metal near the magnet was copper, the needle of the galvanometer was thrown 80°, 90°, or more, from its natural position. Faraday was unable to obtain chemical effects by the induced current, but on repeating his experiments with an armed loadstone capable of lifting about thirty pounds, he succeeded in convulsing powerfully the limbs of a frog, and in obtaining physiological effects upon himself.

An important element in magneto-electric induction, which was noticed by Faraday during the prosecution of his earliest experiments, is *time*. Volta-electric induction is sudden and instantaneous, but magneto-induction requires sensible time, and experiment proves that an electro-magnet does not rise to its fullest intensity in an instant. Fig. 4 shows a convenient arrangement for exhibiting and illustrating magneto-electric induction. Two or three hundred feet of cotton or silk-covered copper wire are wound round a hollow pasteboard or wooden cylinder, and the ends connected with a galvanometer placed at a distance. On thrusting a tolerably powerful bar magnet into the axis of the cylinder, the needle is immediately and strongly deflected; on allowing the bar to remain at rest, it soon regains its natural position, but is again deflected in an

Fig. 4.

opposite direction when the magnet is suddenly withdrawn; the motions of the needle are reversed when the opposite end of the magnet is thrust into the cylinder.

Faraday's great discovery of magneto-electric induction has been utilized in a variety of ways. Pixii, of Paris, was the first (in 1832) to make public a machine by which a continuous and rapid succession of sparks could be obtained from a permanent magnet. In the following year his machine was much improved by Saxton, who exhibited in the Gallery of Practical Science, which at that time existed in Adelaide Street, Strand, an instrument by which platinum and iron wires could be fused; chemical decomposition energetically effected; soft iron magnetized; and powerful shocks given. Improvements continued to be made, until in 1842 a patent was secured by Mr. Woolwich, of Birmingham, for the application of the magneto-electrical machine to the art of electro-plating. A full and illustrated account of this ingenious apparatus will be found in the "Mechanics' Magazine, "Vol. xxxviii., p. 146. In the celebrated "depositing" works of Messrs. Elkington, of Birmingham, a large magneto-electric machine is in continuous operation; it deposits 17 oz. of silver per hour.

To Mr. Holmes is due the important application of magneto-electricity to the production of electric light. His machine must be familiar to all who visited the International Exhibition of 1862. The currents are induced by the rapid passage of coils of copper wire wound round soft iron cores between the poles of powerful horse-shoe magnets. The alternately inverted currents thus produced are transmitted by means of a commutator in one direction only, through the carbon electrodes of an electric lamp, likewise his invention, and an extremely brilliant, regular, and constant light, without flashes, is produced. A somewhat similar machine had previously been constructed by Mr. Shepard, for the purpose of producing illuminating gas by the decomposition of water. Holmes's machine and lamp has been in successful operation at Dungeness lighthouse since June, 1862.

The applications which have been made of magneto-electric machines to the administration of shocks for medical purposes are very numerous; none are more convenient, economical, or effective than the arrangement of Mr. Ladd, shown in Fig. 5. The force of the shocks is regulated simply and accurately

by an arrangement by which the distance of the
armature from the poles of the magnet can be
increased or diminished, and the noise which in the

FIG. 5.

earlier machines was produced by the working
together of the metallic cog wheels, has been obvi-
ated by the substitution of discs of vulcanized
rubber.

Another important application of magneto-electricity is the explosion of mines and submarine gunpowder, or gun-cotton charges, in military or
engineering operations. A very compact arrangement of magnets was devised for this purpose by
Mr. Wheatstone, by which an extremely rapid
succession of currents could be established in such
a manner that the effect was almost equal to a continuous current, and he has since effected further
important improvements in the "magnetic exploder"
whereby its size and weight have been very considerably diminished. A remarkably sensitive com-

position for the priming material used in the fuze has been discovered by Professor Abel, and if the fuzes so charged are arranged in the branches of a divided circuit, their ignition in numbers varying from two to twenty-five is certain; the only important precautions to which it is necessary to attend rigidly in order to ensure uniform access are the proper insulation throughout of the main wire and branch-wires leading from the instrument to the charges, and the thorough protection of all connections of wires from access of moisture.

The magnetic exploder has important advantages over the application of the voltaic battery for firing charges, the principal being — its constant fitness for use, its portability, and its small liability to derangement. The magnet fuze is, moreover, more certain than any fuze arrangement applied with voltaic batteries. It may be preserved for a great length of time in any

Fig. 6.

climate, and will bear very rough treatment without chance of injury.

The latest improvement in the construction of the magnetic exploder is shown in Fig. 6.

9.—INDUCTION OF A CURRENT ON ITSELF.

In his ninth series of Experimental Researches, read before the Royal Society, Jan. 29th, 1835, Faraday makes known a new action of the electric current, viz., an induction on itself. The inquiry arose out of a fact communicated by Mr. Jenkin, which is as follows:—If an ordinary wire of short length be used as the medium of communication between the two plates of an electromotor, consisting of a single pair of metals, no management will enable the experimenter to obtain an electric shock from this wire; but if the wire which surrounds an electro-magnet be used, a shock is felt each time the contact with the electromotor is broken, provided the ends of the wire be grasped one in each hand; a bright spark also occurs at the place of disjunction. The same effects occur, though by no means in the same high degree, when a simple helix of copper wire is used to connect the opposite poles of the battery without any coil of iron. Thus on connecting one end of a helix of eighty or a hundred feet of stout copper wire with one plate of the battery, and making and breaking contact between the other end of the helix and the opposite pole of the battery, which is best done by dipping it into, and quickly withdrawing it from, a cup of mercury in good metallic connection with the battery, a distinct shock is experienced every time the wire leaves the mercury, although none can be perceived when it enters the fluid metal. If the same quantity of wire be used, not in the form of a helix, the spark is much less bright on breaking contact, nor can any, or only very feeble, physiological effects be obtained; the phenomenon is best observed by taking about 100 feet of copper wire and winding 50 feet into a helix, then bending it in the middle so as to form a double termination, and allowing the other 50 feet to remain extended; on using each half of the wire alternately as the connecting wire of an electromotor, the helix will be found to give by far the brightest spark; it even gives a brighter spark than when it is used in conjunction with one extended half of the wire. The brightness of the spark with the helix is however no proof that more electricity, or electricity of higher intensity, is passing through it than through the extended wire connecting the plates of the battery. Faraday found the effect at the galvanometer and the electrolyzing power to be the same, whether he used a short wire, a long wire, a helix, or an electromagnet; under certain circumstances, however, proof can be obtained that there is a diminution of the battery current when a long wire is used, in consequence of the resistance it sets up; thus, on soldering an inch or two of fine platinum wire to one end of a long wire, and also a similar length to one end of a short wire, and then using each to connect the plates of the same electromotor, Faraday found that the latter became ignited, though producing but a feeble spark, while the former remained cold, though giving a bright spark. This bright spark is due to a momentary *extra* current induced in the wire at the moment the primary or inducing current from the battery ceases to flow through the wire. This extra current, or wave of electricity, may be examined as to direction and intensity by a very simple expedient; thus, by using an electro-magnet, and connecting the ends of the principal wire by a short cross-wire, the bright spark ceases to appear in the mercury cup on breaking battery contact, because the extra current which would have produced it now passes along the connecting wire. If the cross wire be divided in the middle, a spark may be obtained by rubbing the two ends together, while contact is broken and renewed between one of the principal terminal wires, and one of the mercury cups of the battery; chemical decomposition may also be effected there, and a fine platinum wire ignited. The direction of the extra-current is shown by introducing a galvanometer between the divided ends of the cross

B

wire ; whilst the current from the battery is circulating the helix of the electro-magnet, the galvanometer is affected in the direction of the battery
current, because the cross-wire carries one part of
the electricity excited by the battery ; but, if the
needle be forced back to its normal position, and
secured there by pins, and if battery contact be then
broken, the needle is powerfully deflected in an
opposite direction, thus proving that the wave of
induced electricity—the extra current—moves in the
wire, at the moment of disruption with the battery,
in a direction contrary to that of the electrical
current set in motion by the battery itself.

10.—EXTRA-CURRENT.

This extra current may be removed from the wire
carrying the original current, to a neighbouring wire ;
if two helices be arranged on the same hollow pasteboard or wooden cylinder, in close proximity, but
nowhere actually touching, and one used for making
and breaking contact with a battery, the usual bright
spark appears at the moment of disruption ; but if
the two ends of the second helix be brought into
contact, so as to form an endless wire, the spark becomes scarcely sensible, and all the phenomena
described (9) as occurring between the divided ends
of the cross wire may be re-produced between the
two extremities of the second helix. " The case,"
therefore, says Faraday, " of the bright spark shock
on disjunction may now be stated thus :—If a current
be established in a wire, and another wire, forming
a complete circuit, be placed parallel to the first, at
the moment the current in the first is stopped,
it induces a current in the same direction in
the second, the first exhibiting then but a feeble
spark ; but if the second wire be away, disjunction
of the first wire induces a current in itself in the

same direction, producing a strong spark. The strong spark in the single long wire or helix, at the moment of disjunction, is therefore the equivalent of the current which would be produced in a neighbouring wire if such a second current were permitted.'' A brighter spark, then, is produced at the moment of disruption of a long wire joining the plates of a battery than of a short wire, because, though it carries less electricity, it induces on itself a more powerful wave current; if wound round into a coil, the spark is still brighter, because of the mutual inductive action of the convolutions, each aiding its neighbour; and the brightnes of the spark is exalted still higher when the coil encloses a bar of soft iron, because the bar, losing its magnetism at the moment of disruption, tends to produce an electric current in the wire around it, in conformity with that which the cessation of the current in the helix itself also tends to produce. It is not so easy to demonstrate the induction of a wave current at the moment of making contact; by using certain expedients however Faraday succeeded in doing so; thus, when a galvanometer was introduced between the ends of the cross wire, a part of the current from the battery was diverted through it; when the needle had taken up its position it was retained there by pins, contact was then broken, but the needle was prevented from obeying the impulse which the reverse wave would have given to it by the stops; contact being now again made, the needle immediately moved onwards, showing, by a temporary excess of current in the cross communication, a temporary retardation in the helix.

11.—INDUCTION OF SECONDARY CURRENTS AT A DISTANCE.

On the 2nd of November, 1838, a memoir on

electro-magnetic induction was read at the meeting of the American Philosophical Society, by Dr. Joseph Henry, professor of natural philosophy in the College of New Jersey, Princeton. This ingenious electrician employed in his experiments flat coils of insulated copper riband, and helices and spools of fine-covered copper wire; with electricity of low intensity, as from a single pair of plates, he obtained with a flat riband coil 93 feet long brilliant deflagrations and loud snaps from a surface of mercury, but no shocks; but when the length of the riband was increased to 300 feet he obtained strong shocks but less brilliant sparks; with electricity of higher intensity as from a series of pairs of plates, the action of the riband was decreased, but when the current from ten pairs was sent through a spool of wire one-sixteenth of an inch in diameter and five miles long, the induced shock was too severe to be taken through the body, though the spark was feeble; a shock was indeed passed through twenty-six persons at once from this spool, when a battery consisting of six pieces of copper bell wire and corresponding pieces of zinc wire, only one-and-a-half inch long, was employed; nevertheless, when a single pair of plates exposing one square foot and three-quarters of zinc surface was used, scarcely any physiological effects could be obtained. In these experiments, contact with the battery was broken and renewed by drawing one end of the riband or helix across a rasp which was kept in good metallic contact with one of the plates of the battery. When the current from a small battery was sent through a copper riband on the top of which was placed a helix containing 3,000 yards of covered copper wire 0·02 inch in diameter, a plate of glass being interposed between the riband and the helix, powerful shocks were obtained from the latter as often as the current through the former was interrupted; when the helix was removed and a copper riband sixty feet long substituted for it, very feeble shocks could be obtained; but sparks were produced; on rubbing the ends of the riband together, needles were magnetized, temporary magnetism in soft iron was developed, and water was decomposed; none of these latter effects could, however, be obtained with the helix. Intense shocks, and magnetizing, and chemical effects also were obtained from thé five-miles spool of wire, when the riband was opened so as to receive it in the centre, and an interrupted current from a single pair of plates sent through the riband.

From these experiments it will be seen that the induced or secondary current obtained from ribands or short stout copper wire partakes of the character of what is called *quantity*, while that from great lengths of fine wire exhibits the qualities of what is termed *intensity*. When the current from an extensive series of plates was sent through a riband and intermitted, no secondary effects could be obtained in the helix, but when the same battery was used with a helix, powerful shocks were induced in a second helix, and sparks and magnetizing effects obtained with a riband; "hence," observes Henry, "an intensity current can induce one of quantity, and a quantity current can induce one of intensity." The induction of a secondary current at a distance from the primary was illustrated by Dr. Henry in a surprising manner, by coiling the wire of the five-mile spool into a ring four feet in diameter, and placing parallel to it another ring of copper riband 270 feet long; on sending an intermitting current from a single pair of plates, zinc surface 35 feet, through the latter, shocks could be obtained from the former at a distance of four feet, and at a distance of twelve inches they were too severe to be taken through the body.

12.—INDUCED CURRENTS OF THE THIRD, FOURTH, AND FIFTH ORDER.

Professor Henry was the first to show that the induction of electricity does not stop with the production of secondary currents, but that currents of a third, fourth, and even of a fifth order may be obtained. An intermittent current from a single pair of plates was sent through a copper riband, a second riband being placed over it to receive the induced secondary current. The ends of this second riband were connected with the ends of a third placed at a distance, and over this a helix of 1,660 feet of fine wire. On grasping copper handles metallically connected with the ends of this helix, powerful shocks were obtained ; thus the secondary current produced a new induced current in a third conductor. By a similar arrangement shocks were received from currents of a fourth and fifth order, and with a more powerful primary current and additional coils, a still greater number of successive inductions might be obtained. The arrangement of coils and helices is shown in Fig. 7, where *a* represents a cylindrical

FIG. 7.

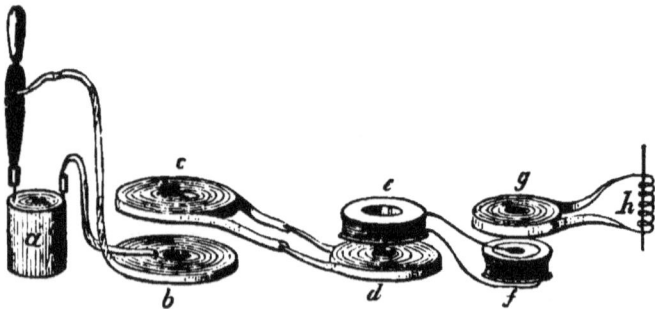

copper and zinc single-cell battery; *b*, a coil of copper riband, about 100 feet long, and an inch and a half

wide ; *c* and *d* similar ribands, about 60 feet long ;
e, a helix of 1,660 yards of copper wire, one forty-
ninth of an inch in diameter ; *f*, a helix of about
1,200 yards of the same wire ; *g*, a copper riband, 60
feet long, and three-quarters of an inch wide ; and
h, a cylinder of about thirty spires of copper wire,
so small as just to admit a sewing needle in its axis.
Now here, as with a primary current only, it is found
that a quantity current can be induced from one of
intensity, and the converse ; for the induction from
coil, *b*, to helix, *e*, produces an intensity current, and
from helix, *f*, to coil, *g*, a quantity one, as is demon-
strated by the magnetization of the steel needle in
the copper spiral *h*. Then, as to the direction of
these induced currents, it was found that there
exists an alteration in the direction of the several
orders, commencing with the secondary, as follows :—

Primary current . . . . . . +
Secondary current . . . . . +
Current of the third order . . —
Current of the fourth order . . +
Current of the fifth order . . . —

the directions being determined by the nature of the
polarity of the magnetized needle, by decomposition,
and by the galvanometer. Induced currents of the
different orders are also produced from ordinary electricity. On discharging a large Leyden phial
through a spiral of tinfoil pasted round a glass
cylinder, a similar spiral of foil being pasted inside
the cylinder, the ends of which were connected with
a magnetizing spiral enclosing a steel needle, the
latter was magnetized in such a manner as to indi-
cate an induced current through the inner riband in
the same direction as that of the current of the jar ;
a spark was also produced when the ends of the
spiral were separated by a small interval. Induced

currents of a third and fourth order were obtained when a large Leyden phial was substituted for the battery, the coils being furnished with a double coating of silk, and the conductors separated by a plate of glass. . By using a powerful Leyden battery, Dr. Henry obtained evidence of the induction of a secondary current at the surprising distance of twelve feet. This subject has more recently been investigated by Reiss, who found that the currents of the third, fifth, and other *odd* orders have the same direction as the original current, and those of the second, fourth, and other *even* orders, have among themselves one and the same direction.

13.—THE ELECTRO-MAGNETIC COIL MACHINE.

We are now prepared to understand the *modus operandi* of those arrangements, which in so many forms have been rendered familiar to the scientific world under the general name of the *Electro-magnetic Coil*. Various as are the external appearances of these machines, they are all based on Faraday's great discoveries of electric and magnetic induction. The first induction coil without an iron core is described by Faraday in par. 6 of the first series of the "Experimental Researches" (Nov. 24th, 1831); the first with an iron core in par. 34 of the same series. By these the important discoveries of electric and magnetic induction were made, and they were carried forward to higher conditions in the ninth series (Dec. 18th, 1834), pars. 1,053, 1,063, 1,090, 1,095, in relation to the action of an electric current upon itself. Having discovered the power, Faraday abstained from proceeding to exalt it. He says, par. 159, second series, "I have rather been desirous of discovering new facts and new relations dependent on magneto-electric induction than of exalting the

force of those already obtained, being assured that
the latter would find their full development hereafter "; and again, in par. 1,118, ninth series (Dec.
8th, 1834), "In the wire of the helix of magneto-
electric machines an important influence of these
principles of action (the inductive action of a current)
is evidently shown. From the construction of the
apparatus, the current is permitted to move in a
complete metallic circuit of great length during the
first instants of its formation ; it gradually rises in
strength, and is then stopped by the breaking of the
metallic circuit, and thus great intensity is given *by
induction* to the electricity which at that moment
passes (see pars. 1,064 and 1060 of the same series).
This intensity is not only shown by the brilliancy of
the spark and the strength of the shock, but also by
the necessity which has been experienced of well
insulating the convolutions of the helix in which the
current is formed ; and it gives to the current a force
at these moments very far above that which the
apparatus could produce if the principles which
form the subject of this paper were not called into
play."

The anticipations of Faraday that his discoveries
would find their "full development hereafter" were
not long in being realized. One of the first electro-
magnetic coils which obtained public notoriety was
that designed by Professor Callan, of Maynooth
College.

It consisted of a coil of thick, insulated, copper
bell-wire, wound on a small bobbin, to serve as the
primary coil, and of a coil of about 1,500 feet of thin
insulated wire wound round a large cylinder, into the
axis of which the smaller coil could be introduced, to
act as the secondary. The ends of each coil are
attached to binding screws, to establish, on the one
hand, a communication between the primary coil and

the battery, and, on the other, for the convenience of interposing any apparatus on which the effects of the secondary current are to be tested. Various contrivances have been adopted for breaking and renewing battery contact, some of an automatic character, others requiring manual assistance. Dr. Bird was the first in this country, at least, to employ the permanent magnet to effect rupture of contact ; this he did by causing a small bar electro-magnet to vibrate between the opposite poles of a pair of steel horse-shoe magnets in such a manner that every time each arm of the electro-magnetic bar rose and fell it should effect a disruption and a renewal of contact between the battery and the primary coil; in this way he obtained 300 oscillations in a minute, and a series of induced currents, capable not only of communicating intense shocks, but of exerting powerful electrolytic action ; when a bundle of soft iron wires was introduced into the axis of the primary, the physiological and chemical effects were greatly exalted; the sparks at the contact-breaker were much increased in brilliancy, and were accompanied by a loud snapping noise and a vivid combustion of the mercury. In other arrangements the coil is placed vertically, and battery contact broken and renewed by the rotation of a soft iron bar, mounted between two brass pillars, and situated immediately over the axis of the coil, in which is placed a bundle of iron wires ; in others, a small disc of iron is kept vibrating, with amazing rapidity, over the bundle of iron wires, contact being broken and renewed between surfaces of platinum, which dispenses with the use of mercury. Mr. Henley, some years ago, made and presented to the writer a very powerful electro-magnetic machine, consisting of a series of U-shaped bars of soft iron, round which were wound four coils of No. 34 wire. Contact was broken and renewed by mercury. With this instrument a secondary spark could be obtained passing one-eighth of an inch through air; by a very simple contrivance the ends of the secondary coil could be united and disunited by merely turning an ivory knob; the instrument is, therefore, well adapted for demonstrating the inductions and reactions of electrical currents; when the ends of the secondary are disunited the sparks of the primary are large and brilliant; when united, they are small and faint. But the secondary coil may be dispensed with altogether, and this is one of the best arrangements when the instrument is to be used for medical purposes. The writer constructed a machine of this kind some years ago, containing 100 yards of covered bell-wire surrounding a core of iron wires, battery contact being broken and renewed by clock-work, so that the frequency of the induced shocks could be regulated with the greatest regularity and precision ; wires leading from either end of the coil, and attached to suitable binding screws on the stand of the apparatus, served to convey the extra current, in accordance with the principles laid down in 9 and 10. The frequency of the shocks was regulated by the clock-work mechanism, and the intensity by a water regulator—which ingenious and useful appendage to the medical coil was the invention of the Rev. F. Lockey. This was included in the circuit of coil, and by increasing or diminishing the distance between the wires, so as to interpose a greater or less thickness of water, the power of the shocks could be modified to any required degree, giving the operator such perfect command over the instrument as to enable him to apply this form of electricity to as delicate an organ as the eye, or to administer powerful shocks.

It has recently been discovered by Mr. Wilde,

Professor Royal Society, April 26th, 1866, that an indefinitely small quantity of magnetism is capable of evolving an indefinitely large amount of dynamic electricity. When the wires forming the polar terminals of a magneto-electric machine of peculiar construction were connected for a short time with those of a very large electro-magnet, a bright spark could be obtained from the helices of the latter twenty-five seconds after all connection with the magneto-electric machine had been broken; hence it would appear that the electro-magnet possesses the power of accumulating and retaining a charge of electricity in a manner somewhat analogous to that of the Leyden jar. Mr. Wilde also noticed that the helices of the electro-magnet opposed a certain resistance to the magneto-electric current, and that it required in some cases nearly half a minute before the current attained a permanent degree of intensity. Four permanent magnets which collectively could only sustain 40 lbs., could be made to evolve an amount of electricity sufficient to excite an electro-magnet to such a degree as to enable it to sustain 1,080 lbs., and by suitable arrangement the electro-magnet could be made to evolve a large amount of dynamic electricity. The magneto-electric current produced by a machine containing six permanent magnets, which weighed only 1 lb. each, and collectively could only sustain 60 lbs., was made by Mr. Wilde instrumental in producing a prodigious amount of dynamic electricity. The direct current from the magneto-machine was sent through the coils of the electro-magnet of an electro-magnetic machine, and the direct current from the latter was sent through the coils of the electro-magnet of another much larger electro-magnetic machine. The result was the production of an amount of magnetism in the latter far exceeding anything that has hitherto been produced, accompanied by the evolution of an amount of dynamic electricity so enormous as to melt pieces of cylindrical iron rod, 15 inches in length and one quarter of an inch in diameter, and to produce in the electric lamp a light which cast the shadows from the flames of the street lamps a quarter of a mile distant upon the neighbouring walls, and threw rays from the reflector having all the rich effulgence of sunshine. The light and heat are increased according to the amount of mechanical force employed.

Fig. 8 represents the permanent magnetic arrangement of the above machine, and consists of sixteen magnets, each weighing 3 lbs. The armature is

FIG. 8.

rotated by a multiplying wheel arrangement turned by hand. The current obtained by this means is sufficient to heat to whiteness 5 inches of platinum wire, ·012 in. diameter, and with one of Mr. Ladd's

Inductoria containing three miles of secondary wire 2-in. sparks can be obtained. The commutator can be arranged to send the currents in one direction, and will then liberate from acidulated water one-and-a-half cubic inches of the gases per minute. It can also be used for the various lecture experiments, where a battery has hitherto been indispensable. It is well adapted for blasting purposes, and likely to be extensively used for electro-plating, etc.

14.—ELECTRO-MAGNETIC COILS FOR THE MEDICAL ADMINISTRATION OF ELECTRICITY.

At the International Exhibition of 1862, a great number of induction coils for medical purposes were exhibited; they were generally arranged with much ingenuity, and with varied means of altering the number of shocks per minute, as well as of the strength of the shocks.

Mr. Ladd's very convenient arrangement of the electro-magnetic medical coil is shown in Fig. 9. The electromotor is a sulphate of mercury battery, which has been chosen for its extreme cleanliness, and high electro-motive force. The apparatus when closed resembles a small book with a clasp, and is very portable. On the left-hand side of the book is a small door, which, upon being opened, exposes the sulphate of mercury battery. The tray is made of ebonite; within this is a cell of carbon cut out of a solid block; this is lined with a piece of cloth or lint, and upon this is placed a slab of zinc, a piece of which is bent up, and faced with platinum; there is also a copper connecting piece for the carbon cell; on the right-hand side of the tray, the poles, vibrating spring, etc., are placed.

To excite the battery, a sufficient quantity of sulphate of mercury is placed on the carbon tray to cover it over and make an even surface; the lint is placed above this and left sufficiently large to turn

FIG. 9.

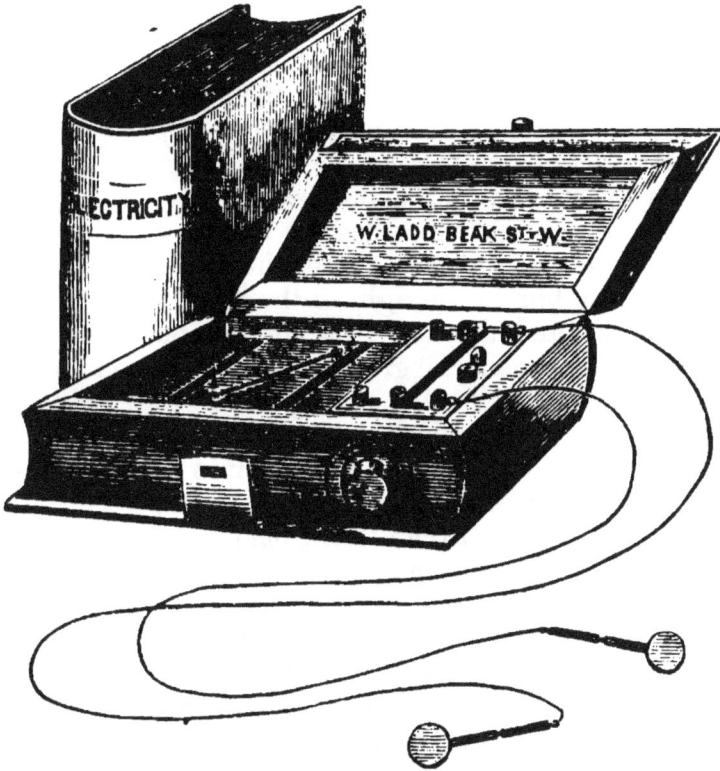

up at the sides, so as to prevent contact between the zinc and carbon; about a teaspoonful of water is then poured on it and the zinc plate placed on the lint; the tray is now put back into the box and closed. The battery is now in circuit with the primary wire of the coil; the spring must next be adjusted by the eccentric button, which must be gently turned round until the vibrations show that the battery is in action; by turning the button back a little, the vibrations are diminished in frequency. On either side of the vibrating spring will be perceived two nuts with holes through them, those on the left marked P+ and P—, those on the right S+ and S—. P+ means the positive pole of the primary wire. P— the negative pole of the primary wire. S+ signifies the positive pole of the secondary or finer wire ; S— the negative pole of the same wire. If a very gentle current be desired, the copper pegs of the conducting wires must be inserted into P+ and P— respectively, and upon holding the conductors in the hands, the physiological effects are scarcely perceptible; to increase these, the brass handle in front of the box to the right of the clasp is gradually drawn out, and the soft iron core contained in the centre of the coils is gradually exposed and mag- netized, increasing the strength of the induced current. On pushing back the brass tube and inserting the pegs of the conducting wires into the nuts S+ and S— the current from the secondary wire, which is far more powerful than that from the primary wire, is obtained. If now, magnetic induction be added to volta-induction by gradually drawing out the brass tube, the current becomes by degrees so powerful as to be unbearable; thus, with this little battery, any *requisite* amount of power may be obtained.

If it is in regular daily use, the carbon cell will have to be cleaned about once a week; the lint should be taken out and well washed, so as to remove all the yellow deposit; the carbon cell then rinsed out with fresh water, and the under surface of the zinc well washed ; the lint is now replaced, and the battery is ready to be re-excited; the process of cleaning need not take more than two or three minutes.

The form of voltaic battery in which sulphate of mercury and carbon electrodes are substituted for the sulphate of copper and copper electrodes of the

Daniell's battery, is known as the "*pile Marie Davy*," an arrangement of many hundred cells has been constructed by Mr. Gassiot, with which he now exhibits some of his most beautiful and striking experiments on electrical discharge through various vacua. The Marie Davy battery has not the power of that of Daniell, but it is clean and remarkably constant.

15.—THE INDUCTION COIL.

Up to about the year 1842, the only object sought by makers of electro-magnetic machines would seem to have been the production of shocks, and the regulation of their intensity and frequency. It was M. Masson who first directed attention to other static phenomena which the instrument was capable of developing; in that year he constructed, in conjunction with M. Breguet, an apparatus with which, though consisting of a single coil only, and that very imperfectly insulated, he was able to obtain sparks in rarefied air of sufficient length to show the unequal heating powers of the two poles of the circuit; to charge a condenser, and to ignite platinum wire; these electricians were therefore the first to show that, by the process of induction, the electricity of the galvanic battery (dynamic) is converted into the electricity of the ordinary electrical machine (static).

In 1851, M. Ruhmkorff, an intelligent and well-known philosophical instrument maker in Paris, directed his particular attention to the more perfect insulation of the wire, which, after covering in the usual way with silk, he surrounded with a coating of gum-lac, and attached the ends to glass rods, rightly concluding that the wooden frame of the instrument, though sufficiently insulating for voltaic,

c

was not so for static electricity. He moreover diminished the diameter of the coil, thereby, with the same quantity of wire, obtaining a greater number of convolutions; and he greatly increased the length of the secondary, extending it in some of his machines to the length of nearly six miles. Lastly, from a conviction that the magnetic current was more effectual in arousing an induced current than the mere coil, that is, that the secondary effects were referrible more to *magneto-* than to *volta*-electric induction, he gave in his coils a great development to the former, by introducing into the axis of the primary a large bundle of iron wires, which he found to acquire a much higher degree of magnetism than an equal weight of iron in the form of an iron bar. To interrupt the inducing current, he employed a simple piece of mechanism known as "Neef's" hammer, consisting of a small block of iron, which vibrated between the projecting end of the coil of iron wires and a small anvil connected with the primary coil, in such a way that when the anvil and hammer were in contact the current was *on*, but the moment they separated it was *off*. It will be unnecessary to describe minutely this form of contact-breaker, as it has given place to other and far more efficient arrangements. With these improvements Ruhmkorff obtained effects which were at that time surprising; he not only got brilliant sparks between the terminals of the secondary wire, but between the wire itself and a body out of the circuit in communication with the earth; and he obtained a discharge, in a vacuous globe, of great brilliancy, the spark filling the balloon with that magnificent phenomenon, stratified light, about which we shall have more to say presently. These effects were greatly exalted in degree, by interposing in the circuit of the primary, a simple condenser, as recommended by M. Fizeau;

brilliant and crepitating sparks in free air were now obtained, three-quarters of an inch long, and the shock was so violent, that it is stated by Du Moncel, that M. Quet, incautiously getting himself into the circuit, was knocked down, and so much injured as to be obliged to keep his bed for some time, nevertheless the battery only consisted of six elements. We are reminded by this story of the account given by Muschenbroek of the effects on himself of his first shock from a Leyden phial, which he declared deprived him of his breath, and made him ill for two days; it is, however, true that great care is necessary in experimenting with the induction coil as at present constructed, as an incautious contact with the secondary wire communicates a most disagreeable shock; though how M. Quet came to be so much affected, unless he wantonly placed himself directly in the circuit, we are at a loss to understand. Various forms are given by M. Ruhmkorff to his coil; the bobbin is sometimes arranged vertically, though generally horizontally, and the ends are backed up and supported by discs of glass or gutta percha, through which the wires of the secondary pass to their insulating pillars. The size of the primary wire is about 0·078 in. in diameter; the secondary wire is the No. 28 of commerce; and the instrument is furnished with a commutator, for the purpose of reversing at will the direction of the current.

Shortly after Ruhmkorff's improvements were announced, Mr. Hearder exhibited one of his improved machines at the Royal Cornwall Polytechnic Society; it was six inches in length, and contained about a mile and a half of fine secondary wire; it was wound upon a hollow bobbin of wood, covered with gutta-percha, and having its centre large enough to contain the primary coil with its iron core. The secondary wire was covered with silk, and the layers insulated from each other with oiled silk and gutta-percha; it was provided with a condenser, gave sparks between the terminals more than ·one inch in length, and charged a Leyden jar containing three square feet of surface, so as to give a torrent of brilliant discharges between platinum terminals. For this instrument Mr. Hearder received the Society's first silver medal.

In September, 1856, Mr. Charles Bentley showed the writer a coil of his own construction, which gave sparks between terminals of silk-covered wire an inch and a half long, the primary being excited by five of Grove's cells. In building up this coil he used, as an axis, a hollow iron tube, nine or ten inches in length and half an inch in diameter; round this he arranged a considerable number of insulated wires, the same length as the tube, and sufficiently numerous to form a bundle of an inch and three-quarters in diameter. This core was insulated by being covered with six or eight layers of waxed silk. Thirty yards of No. 14 cotton-covered copper wire were then wound carefully round the iron core, forming two layers, which were then insulated from each other by eight thicknesses of waxed silk. The secondary wire consisted of 3,000 yards of No. 35 silk-covered copper wire, and the coils which it formed were insulated by several layers of gutta-percha tissue; it was wound so as to leave a space of about one-sixteenth of an inch at either end of the coil beneath, so that it formed a cylinder with rounded ends — a form preferred, from its obviating the necessity of glass checks for keeping the wire in its place. The condenser, which was contained in a separate box, consisted of 100 sheets of tinfoil, 4 × 9 inches, each sheet of foil being placed between two sheets of carefully-varnished paper, and the alternate ends connected with appropriate binding screws.

The induction coil, as now constructed by Mr. Ladd, which is shown in Fig. 10, and more conspicuously in the frontispiece, consists of the usual 'primary,' which is of covered copper wire, ·10 inch in diameter, or No. 12 wire gauge, wound into a coil of three thicknesses, enclosing a bundle of iron wires 1·8 inches in diameter; the ends of this fasciculus project ·7 inch beyond the gutta-percha ends, which are seven inches in diameter and ·6 inches thick; these gutta-percha discs are firmly fixed on the base-

FIG. 10.

board of the machine, and serve both to support and to insulate the coil. The secondary' is a coil of No. 35 silk-covered wire, three miles long; it is very carefully wound round the primary in about thirty layers, each layer insulated from its neighbour by a sheet of gutta-percha. The total length of the coil is eleven inches, and its diameter, including the velvet jacket, five inches. The ends of the secondary pass through one of the terminal gutta-percha discs to an insulated discharger, the arms of which move in ball-and-socket joints, so that the terminals may be separated any distance from one another up to about four and a half inches. The arm in connection with the wire proceeding from the *interior* of the coil is provided with an ivory handle, with which the arm may be moved; the other arm, in connection with the *exterior*, terminates in a brass knob; this must not be touched while the machine is in action, if the operator wishes to avoid a powerful and painful shock. One of the ends of the primary is brought out through the anterior and the other through the posterior gutta-percha disc, to two brass studs, from which they are conducted underneath the wooden base to the commutator and the contact-breaker. The wires from the battery (five pairs of Grove's arrangement, immersed platinum $5\frac{1}{2} \times 3$ inches), are attached to two binding screws, one on either side of the commutator, as shown in the frontispiece. The condenser is conveniently placed in a box underneath the base of the instrument, to which it is firmly attached. It is composed of about fifty sheets of tinfoil, 18×8 inches, and between each sheet is laid a sheet of varnished paper; one-half of the foil is in metallic connection with each side of the break, so that when contact is broken the interrupted ends are respectively in metallic communication with the opposite coatings of the condenser.

The contact-breaker merits especial notice, as it is to the improvements introduced into this part of the apparatus that the surprising effects of the coils of the present day are in a great measure to be ascribed. In Ruhmkorff's original instrument, the interruption of the battery current was, as we have seen, effected by the rising and falling of a small iron hammer; this, whilst it accomplished the general purpose of breaking and renewing battery contact, set up no resistance, the hammer being raised as soon as the iron core had received sufficient magnetism to enable it to attract a very small piece of iron, whilst the falling of the hammer on the

interruption of the current was in no way influenced
by the degree of magnetization of the iron core.
The contact-breaker is now constructed by Mr. Ladd,
with the object of giving the operator the means of
setting up a greater or less resistance to the attractive force exerted by the magnetic iron core. This
is accomplished by attaching the hammer to a stiff
spring, placed vertically, as shown in Fig 11. where
A is the disc of iron capping one end of the iron
core; B, the iron hammer of the contact-breaker,
surmounting a stiff spring attached to a brass stand
screwed to the base-board of the instrument; c is a
little projecting nipple, tipped with platinum; d, a

FIG. 11.

corresponding little disc of platinum, soldered to the
end of a screw, which passes through the top of a
brass pillar, firmly screwed down to the base-board;
the distance between d and c can be regulated with
the greatest nicety by the thumb-screw, e. Now,
when c and d are in contact, and the commutator is
turned on, the battery current is circulating round
the primary coil, the fasciculus of iron wires becomes

a more or less powerful magnet, according to the power of the battery; B is attracted to A, by which act *c* and *d* are separated; battery contact is hereby broken, and the effects of the induced current are obtained at the terminals of the secondary. But if the action of the contact-breaker ended here, it would be nothing more than Neef's hammer placed vertically; it will be seen, however, that by turning the screw *g*, the point *f* attached to its axis may be made to press with greater or less force on the spring supporting the hammer, thereby keeping *c* and *d* more or less firmly in contact, and necessitating a corresponding degree of magnetization of the fasciculus to part the platinum discs; when, however, this has been attained, contact with the battery is instantly broken, and the hammer is forced back with violence by the conjoint action of the spring and screw; *d* and *c* again come into contact, the iron core again becomes magnetic, A attracts B, and the battery current is stopped, *c* is again forced upon *d*, and so on. Now a degree of pressure may be exerted on the spring support of B by the screw *g* sufficiently great entirely to overcome the attractive force of A; under such circumstances the instrument is, of course, passive, but by gradually relaxing the tension to a certain degree, the magnetic power of the core just overcomes the antagonistic force of the spring, and then it is that the most powerful inductive effects are obtained, evidently because then the fasciculus has received from the battery its maximum amount of magnetism, which it loses instantaneously by the interruption of the battery circuit, giving rise to a powerful wave of induced static electricity in the secondary coil. The influence exerted by the resistance thus set up to the rupture of battery contact on the strength of the induced current is far greater than could have been anticipated. The instrument

we have been describing gives between the terminals
of the secondary, when the screw g is entirely relaxed,
thin thready sparks, about $1\frac{1}{2}$ inch long, but when
the spring is strained to the utmost, brilliant flashes
upwards of 4 inches long, pass continuously.
The control which this form of breaker gives to the
operator while performing experiments in which considerable variations in the power of the induced
current are required, renders it of great value.

The Condenser.—The function of this very important part of the modern Induction Coil is by no means
clearly understood. Fizeau, who suggested it, says,
that it condenses and destroys, by a static effect, the
electricity of tension or induction which gives rise to
the extra current in the induction wire, and which
reacts on the induced current in the secondary wire
in a direction contrary to that of the voltaic current.
Faraday seems to have much the same opinion. He
says :—" When the secondary current is interrupted,
the inducing power of the primary current acts in
its own wire to produce certain hurtful or wasteful
results ; the condenser takes up this extra power at
the moment of time, and converts it to a useful final
purpose upon principles belonging to static induction." Poggendorff's view is that the function of
the condenser is to draw away the electricity of
tension which, when the battery current is interrupted, accumulates at the two ends of the inducing
coil, where it would otherwise be retained by the
resistance of the air reacting on the fluid set in
motion in the thin wire, and so diminishing its
intensity. Hearder suggests, that at the moment of
breaking the contact the induced current exhibits its
intensity at the points of separation by overleaping
the interval ; but if these two interrupted ends be in
contact with the extended conductors of the condenser,
a portion of this intensity may possibly be reduced

by its being determined in the direction of the two conductors, which, by inducing upon each other, have their capacities for electrical charge considerably increased, and thereby act as capacious reservoirs, in which these intensities may expand and exhaust themselves. Whatever may be the true explanation of the *modus operandi* of the condenser, it is certain that it increases vastly the static effects of the induced current, although it does not increase the quantity of the electricity set in motion. Mr. Ladd fits up some of his coils with a simple arrangement for detaching the condenser; if this be done while sparks or flashes four inches in length are leaping between the wires of the discharger they immediately cease, and the terminals require to be brought within half an inch of each other before thin thread-like sparks can be made to pass between them.

Since the former edition of this little work was written, the induction coil has received wonderful developments. Some fine instruments were shown at the International Exhibition of 1862. One by Siemens and Halske is especially noticed in the Jurors' report for the great length of spark obtained (from one to two feet in length) with a comparatively very small length of wire in the secondary coil, which is stated to be $6\frac{1}{2}$ miles. A singular mistake, was, however, made in stating the length of the secondary coil—instead of being 10,755 metres, about $6\frac{1}{2}$ miles, it was in reality 129,000 metres, or nearly 80 miles, so that the instrument was in no way remarkable for power. An admirable coil was constructed for Mr. Gassiot by Mr. Ritchie, a philosophical instrument maker, of Boston, U.S. The primary wire is wound in three courses on a helix of 150 feet in length. The secondary helix is divided into three bundles, each 5 inches long, wound on cylinders of gutta-percha, the upper and lower coils are each 25,575 feet in length, and the middle 22,500 feet. The maximum effect with the three coils is to produce a spark 13 inches long ; with five cells of Grove's battery, Mr. Gassiot obtains sparks or flashes $12\frac{1}{4}$ inches long.

Ruhmkorff now constructs coils containing 100,000 metres of wire in the secondary. The writer has had the pleasure of witnessing some experiments with one of these magnificent instruments, the property of Mr. Atkinson. When excited by a single cell of the carbon nitric acid battery (Bunsen's), sparks $3\frac{1}{4}$ inches in length are obtained ; two cells give sparks $6\frac{1}{2}$ inches long ; three cells, sparks $10\frac{1}{2}$ inches ; four cells, $12\frac{1}{2}$ inches ; five cells, sparks 14 inches ; six cells, 15 inches ; and seven cells, sparks 16 inches. Beyond this it is hardly safe to go, for fear of injury to the coil ; but sparks or flashes upwards of 19 inches in length have been obtained.

Mr. Ladd constructs (and showed at the International Exhibition) induction coils from which he obtains 5-in. sparks, using five cells of Grove's battery with plates 5 × 3 in. immersed. The construction is as follows :—On a core of iron about a foot long are wound fifty yards of copper wire of No. 12, B.W.G. (0·034 inches) insulated with cotton. This coil forms three layers, round which five or six thin gutta-percha sheets are wrapped. The secondary coil, formed of three miles of No. 35 copper wire (0·005 inches), insulated with unvarnished silk, is wound backwards and forwards along this core with each layer insulated from the preceding one by five or six sheets of thin gutta-percha.

Mr. Ladd also exhibited a coil of very different proportion. The iron core and primary coil are about 8 inches long ; but the secondary coil, placed in the centre of its length, is only 4 inches long, but $7\frac{1}{2}$ inches in diameter. Much the same results may be obtained with this as with the preceding coil.

An induction coil constructed by Mr. Ladd for Dr. Robinson, of Armagh, in which there were two secondary coils—each containing 5,690 yards of wire, together therefore 6 miles 820 yards—gives results which, considering the length of the secondary, are certainly very remarkable. Thus Dr. Robinson writes:—

1 cell gives a spark . 2·04 inches long.
2 cells   ,,   . . 5·06   ,,
3 ,,   ,,   . . 6·45   ,,
4 ,,   ,,   . . 7·65   ,,
5 ,,   ,,   . . 8·38   ,,

The battery cells referred to are Grove's $5\frac{1}{2} \times 4$ in. immersed platinum.

In describing these results, Dr. Robinson remarks:—"There was no internal discharge in any part of the coils. Whilst making these trials, the barometer was at 30·25 inches, which high density of the air from its greater resistance materially shortens the spark : had it been our mean pressure 29·6 in., I feel satisfied that the spark would have been 9 inches."

An end view of one of Ladd's coils is shown in Fig. 12, from which the positions of the contact-breaker a, commutator b, and the binding screws for communicating with the battery, etc., may be seen. The binding screws, $c\,d$, are used for getting the battery spark in connection with the induced magneto-spark and for showing bright scintillations from iron and other metals. The contact-breaker must be firmly united, and the ends of the secondary coil connected. Battery contact is made in the usual way, by the screws on either side of the commutator, and the effects are obtained at the terminals represented by the binding screws, c, d.

In a memoir "On increasing the Electricity given by Induction Machines," recently published, (May 31st, 1866) by Dr. Robinson, some useful practical maxims as to the construction of the *Inductorium* (as

the Germans have named the instrument) are given. By increasing the length of the spark, the object to which the attention of instrument makers has chiefly been directed; no addition to the *quantity* of electricity is made; this is however the most important object, for in most applications of the inductorium, all tensions above what is necessary to force the necessary quantity of current through the circuit is useless, nay, sometimes injurious. Dr. Robinson thinks that a tension which gives sparks of four inches will be found quite sufficient in ordinary cases, and this will be given by about 20,000 spires, all beyond only adding to the weight of the instrument, its cost, and the difficulty of insulation. It must be kept in mind that the mere quantity is independent of the length of wire; it was found actually the same for a flat spiral of twenty-one spires, and for a helix of 13,655.

The quantity increases with the diameter of the wire of the core up to a maximum which is attained when this is about the sixty-fifth of an inch. Helices may be combined either for tension, or quantity, without much loss of their respective powers. If for the former, they are combined in *series*, the general tension is the *sum* of the individual ones, and in this way we can obtain sparks of a length limited only by the strength of the insulator, which is interposed between the primary and secondary helices. If the latter be all of the same wire, the quantity remains unchanged; if they differ in this respect it will be intermediate between the weakest and strongest. If they are combined for quantity, they must be set *collaterally*, *i.e.*, all their positive terminals connected, and their negative. The resulting current will be the sum of all the separate ones, but the *tension* is not increased; the sparks seem even a few hundredths of an inch shorter, but are much denser, and in the higher combinations, approach to the character of a jar discharge, hence there is no risk to the apparatus by extending this mode of combination to any extent.

In combining these instruments, the primaries should not be consecutive if of large numbers, for so, the action of their extra-current (10) would be very destructive to the *rheotome* (contact breaker.) With two primaries containing 726 spires in series, the spark in the mercurial break was almost explosive, but when they were collateral the action was quiet. Were, however, ten or twelve to be so combined, it would require a battery of very large cells to maintain the current, and it is better to have a separate battery for each pair of primaries. The *negative* pole of all the batteries should be connected with the mercury of the rheotome; from its platinum point separate wires must go to the entering bind-screw of each primary; other wires must go from their exit bind-screws to the positive poles of the respective batteries, and thus their action is perfectly synchronous. In this way Dr. Robinson thinks that an amount of electric power which has not hitherto been approached by the inductorium may be obtained.

16.—EFFECTS OF THE INDUCED CURRENT.

In making the following experiments, it is assumed that the operator is working with an instrument such as is figured in the frontispiece, with larger coils the phenomena are of course exalted in a proportionate degree :—

Example 1.—The battery being well-charged—the zinc cells, with a mixture of one part of oil of vitriol and six or eight of water, and the platinum cells with ordinary nitric acid—draw the ends of the discharger about three inches apart, and turn the commutator; brilliant zigzag crepitating flashes will dart between the points, the length of which may be increased to four inches, and sometimes even more, by withdrawing the points gradually (take care *not* to touch the arm which has the brass knob); now bring the points to within about two inches of each other, and observe the spark, it will be found split up into bundles, and to be surrounded with a sort of yellow-green atmosphere, which may be expanded into a mass of irregular violet-coloured flame by gently blowing it. The two parts of the induction-spark, viz., the point of light and the luminous atmosphere, may be completely separated by opposing to one of the electrodes another of a V-shape. By suitably regulating the distance of the extremities of the latter from the former, M. Serrot succeeded in establishing an atmospheric current, which carried the luminous atmosphere towards that branch of the V-shaped electrode which was more remote from the opposite pole. Under these circumstances, the luminous atmosphere appeared only at this latter branch—the other branch receiving the ordinary spark. Du Moncel has also shown that of the two parts of the spark the luminous atmosphere only is affected by the magnet. Dr. P. L. Rijke has made experiments, from which he concludes that the point of light in the inductive spark is to be attributed to the re-compositions of the electric charges accumulated at the extremities of the secondary wire, while the luminous atmosphere is produced by the electric fluid contained in the parts of the wire nearer to its middle point. When the inductive wire is discharged, the electric charges of the two extremities first unite, and the spark is bright, while the charges of the parts nearer the centre, meeting with considerable resistance, require a sensible time, and the spark becomes altered, diminishing in illuminating power and increasing in volume.

Ex. 2.—While the 4-inch sparks are passing, remove the wire which connects the two binding screws on the left-hand side of the base of the instrument (see Fig. 12), thereby disconnecting the condenser ; the sparks will immediately cease, and the wires will have to be brought within a half of an inch before they reappear, now very faint and thin ; re-connect the screws, and the flashes will reappear with their former length and brilliancy. If the spark from an Inductorium be projected on a screen by the electric current, and the impression contrasted with that of the flame of a candle—in the former, two cones are seen to issue from the terminals instead of the single one of the latter, one being more powerful, and overcoming or beating back the other; and this effect is reversed as the direction of the current is reversed. In the voltaic arc there is a transmission of matter, principally from the positive (which is the more intensely heated) to the negative terminal; in the spark from the coil the dispersion is principally, and in some cases appears to be *entirely*, from the negative terminal, which is now the more intensely heated.

Ex. 3.—Attach to the terminals of the discharger, two platinum wires, each about two inches long, and gradually approach them; the wire on the negative side will become intensely heated, and will ultimately fuse; now turn the commutator, thereby changing the direction of the current; the same phenomenon will occur with the other wire; substitute for the platinum wires thin wires of iron, the negative wire will speedily begin to burn with brilliant scintillations; replace the iron by zinc, the negative wire will burn with a brilliant white light. This heating property may be taken advantage of to determine the direction of the induced current. While vigorous sparks are passing between the terminals, introduce a piece of paper, or a thin shaving of wood; either will be speedily kindled.

Ex. 4.—Attach iron filings to a large pane of glass, by means of a suitable varnish, and when dry place it between the terminals; flashes of light more than a foot in length may thus be obtained. Moisten a piece of cork, ten inches long and four inches wide, with dilute sulphuric acid, place the terminals upon it, first about two inches apart; great heat will be set up on the line of discharge, which will vaporize the water, and the cork, becoming charred by the sulphuric acid, will begin to burn; now slowly separate the terminals, drawing one along the surface of the cork, in a zig-zag manner, the flame will

D

follow it, charring the cork in its progress and leaving behind a line of light. In this way you may proceed from one end of the cork to the other, making a complete lake of fire, which has, in the dark, a very beautiful appearance. The best way of making the experiment is to lay the cork upon the table, and stick into one end a wire in connection with the inner terminal of the coil; a wire, leading from the outer terminal, is attached to a brass rod provided with a varnished glass handle, and to this a stout wire; the operator directs the wire along the cork by this contrivance without the chance of getting a shock. If a sheet of silvered leather be substituted for the cork, it becomes brilliantly illuminated with a green-coloured light; or if common leather be moistened with dilute sulphuric acid, it may be used instead of cork. It must be observed that both cork and leather, after having once been rendered conducting by acid, retain their conducting power for a long time after they have become dry.

Ex. 5.—Separate the arms of the discharger beyond the striking distance; in the dark, brushes of light will be seen to dart from the positive electrode, and the negative will be illuminated by a characteristic star of light, also throwing off smaller brushes which re-curve over the wire.

Ex. 6.—In liquids of good conducting power no spark can of course be obtained, but in *non* or imperfectly conducting fluids short crepitating sparks pass. In *oil* these sparks have a greenish white colour; in alcohol they are red and crepitating; in oil of turpentine, and in bi-sulphide of carbon, they are very brilliant. Pour some oil on the surface of water in a glass vessel; introduce a wire covered with gutta-percha, and proceeding from the interior of the coil, underneath the water, just below the oil; and plunge a protected wire from the other extremity within striking distance, in the oil; strong crepitating sparks are obtained, and hydrogen gas is liberated, which burns on the surface of the liquid.

Decomposition of Gaseous Compounds.—When the spark-current from the induction coil is sent through ammonia, it exhibits a violet light, surrounded with a blue edge. At first the mercury over which the gas is confined falls rapidly, the rate of expansion diminishing with the progress of the decomposition; in five minutes the decomposition of a moderate volume of ammonia is accomplished. The original volume is then doubled; the spark current exhibits the pure violet light characteristic of hydrogen, and water injected into the tube produces no diminution of volume. The coil thus becomes a valuable instrument for demonstrating the composition of this interesting gaseous alkali in the lecture room. For the introduction of the spark current through this and other gaseous compounds, the simple apparatus shown in Fig. 13 was contrived by Buff and Hofmann. A fine platinum wire is fused into the shorter limb of a thin U-shaped glass tube, and filed off so as scarcely to project beyond the glass. At a distance of a few millimètres from the platinum pole thus obtained, the loop of a second platinum wire is thrown over the tube, and the wire wound round the tube until it nearly reaches the bend. The tube is then filled with mercury, and the shorter limb introduced into the graduated gas-tube inverted over mercury in a deep cylinder trough. The pole wires of the induction coil being now introduced, the one into the open end of the U-tube filled with mercury, and the other into the mercury of the cylinder trough,

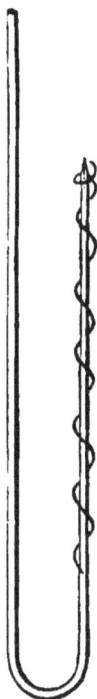

Fɪɢ. 13.

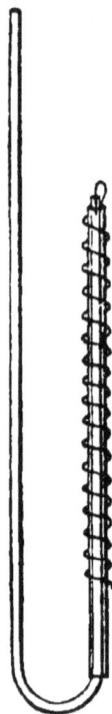

the spark current may be established or interrupted at will, by either depressing the U-tube until the outer platinum wire reaches the mercury surface, or by lifting it so as to break contact. Occasionally Buff and Hofman effected the decompositions by incandescent coils of iron or platinum, or by the electric arc. For experiments of this nature, both limbs of the U-tube remain open. The iron or platinum wire is inserted into the shorter limb, and then coiled downwards round the tube, as shown in Fig, 14. Since the powerful heat emitted from the coil is apt to crack the U-tube, it was found convenient to surround the latter with a somewhat wider glass tube, which separates it from the incandescent coil. The U-tube, as in the previous case, is filled with mercury, and the pole wires of the battery are adjusted in a similar manner. By depressing the U-tube until the lower end of the coil dips into the mercury, the coil may be readily heated ; by raising the end to a proper height above the level of the mercury in the tube, the *arc* may be conveniently adjusted. Amongst the results obtained by these chemists are the following :—*Cyanogen* was not decomposed by the spark-current, but perfectly by electrically incandescent wires, and more rapidly by the electric light, fifty volumes of the gas leaving, after half an hour, fifty volumes of pure nitrogen ; *nitrous oxide* was slowly decomposed by the spark-current into nitrogen and oxygen ; rapidly by incandescent iron, with the formation of sesquioxide of iron and a volume of nitrogen equal to that of the original gas. *Nitric oxide* was decomposed

FIG.14.

slowly by the spark-current, *rapidly* by the incandescent iron coil, the iron burning with splendid scintillations; the residual volume of nitrogen was one-half the original volume of gas. Through dry *carbonic oxide* the spark-current passes with a blue light, but without effect, nor was this gas decomposed either by the incandescent coil or by the electric arc. *Carbonic acid* was decomposed by the spark-current into carbonic oxide and oxygen; the mixture then exploded, reforming carbonic acid; unfortunately the decomposition is too slow for a lecture experiment ; the colour of the spark in the gas is violet. *Marsh gas* was partially decomposed by the spark-current, ten volumes of the gas becoming, in half an hour, eighteen volumes, and the colour of the spark changing from pale blue to violet. *Olefiant gas* was decomposed by the spark-current, which traversed the gas with a pale red light, into carbon and hydrogen; after about twenty minutes, seven volumes of the gas became $12\frac{1}{4}$; had the decomposition been perfect, the volume should have been doubled. *Sulphuretted* and *phosphoretted hydrogen* were both rapidly decomposed by the spark-current, the former with the deposition of sulphur, the latter with that of phosphorus, in the form of a brown powder. These results are sufficient to show what a powerful, elegant, and useful agent of gaseous analysis the Induction Coil is likely to become.

Ex. 7.—Place several lighted spirit-lamps side by side, between the terminals of a universal discharger, connected with ends of the coil; separate the points twelve inches, the sparks will flash through the flames; with a small coil, not capable of giving sparks more than one inch long in cold air, sparks four inches long may easily be obtained through flame.

Ex. 8.—Connect the terminals of the coil with the

inner and outer coatings of a large Leyden phial, and separate the points of the discharger about $\frac{1}{4}$ of an inch, turn on the commutator, whereupon an extremely brilliant discharge will take place between the points, assuming quite the character of the ordinary Leyden discharge; the noise of this continuous discharge is too great to be borne long without discomfort. "I have never," writes Mr. Grove, who first described this magnificent experiment, "witnessed such a torrent of electrical discharges; it is curious to see the absorption, so to speak, of the voltaic power by the Leyden battery. When the maximum effect for a given Leyden jar has been passed, the contact-breaker shows by its sparks the unabsorbed induced electricity, which now appears in the primary wire; an additional jar acts as a safety-valve to the contact-breaker and utilizes the voltaic power, and so on."

With the larger coils, electrical batteries may be charged and discharged with a continuous and almost deafening noise. When a series of Leyden jars are arranged for charging by *cascade*—that is, each jar insulated, the outside of the first in the series connected with the inside of the second, the outside of the second with the inside of the third, and so on, the outside of the last jar being in communication with the earth, the effects produced with Ruhmkorff's 100,000 metre instrument are brilliant in the extreme. A continuous stream of dazzling light, six inches in length, passes between the terminals, accompanied by a roar that cannot long be endured.

By arranging the jar or battery in the manner shown in Fig. 15, a permanent charge may be given to it. The outer coating is brought into communication with one of the poles of the secondary coil, and the inner coating with one of the arms of the universal discharger, the other arm of which is in communication with the other pole of the coil; the points of the discharger are set two or three inches apart. By this arrangement the wave of induced

FIG. 15.

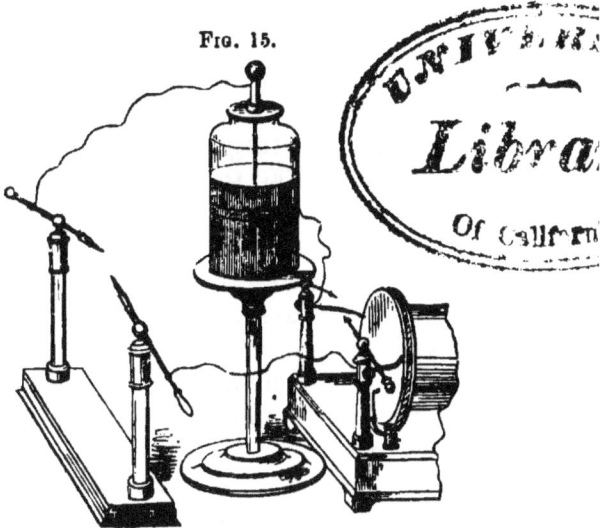

electricity, which is produced as *making* battery contact, is stopped off from the secondary wire; that produced on *breaking* contact, which is by far the most intense, being brought into action. The jar receives therefore a *direct* instead of an alternating charge, and after a few sparks have passed it may be removed and discharged in the usual manner. With Ruhmkorff's large coil a battery containing ten square feet of glass is charged to saturation in a few seconds.

Ex. 9.—Introduce a card between the terminals, arranged as in the last experiment; it will be perforated precisely as with ordinary electricity. Mr. Grove has proposed to count the discharges, by causing a piece of paper to pass with a given velocity per second between the discharging points, and the number of perforations thus made per second may be registered. Mr. Hearder has invented a very ingenious apparatus for carrying out this idea, with which he has endeavoured to compare the effects of the coil with that of an electrical machine, by estimating the amount of glass surface necessary to be rubbed to produce effects equal to those of the coil. The rapidity of the discharges will depend upon the nature of the interrupting spring employed in the coil, and as many as 100 to 200 per second may be obtained.

Ex. 10.—Substitute for the Leyden jar a "fulminating pane," consisting of a square of common window glass, about fifteen inches square, coated on either side with tinfoil; attach to one of the coatings a band of foil, of sufficient length to fold over the edge of the glass and touch the other coating. If this band be wound round a glass rod, the two coatings may be brought within any required distance of each other, by simply winding or unwinding the foil; adjust to the maximum striking distance, and turn the commutator; the discharge now amounts to a positive *roar*, the vividness of the light of which may be appreciated by darkening the room.

Ex. 11.—Fix a piece of platinum wire horizontally across the ball of a Leyden jar, and bring the terminals of the secondary coil respectively near its ends; two interruptions are produced in the secondary circuit, the sparks at which are like each other and equal in quantity of electricity, for the jar as yet forms only an insulating support; now connect either terminal by a wire with the outside of the jar; the spark on that side assumes a bright loud character, but ceases to fire gunpowder, or wood, or paper; and no one would suppose at first, what is the truth, that there is the same electricity passing in one as in the other. The effect of the jar is not to vary the *quantity* of electricity, but the *time* of its passage. That electri-

city, which, moving with comparative slowness
through the great length of the secondary coil, produces a spark having a sensible duration (and,
therefore, in character like that of a Leyden jar
passing through a wet thread) is, when the jar is
used, first employed in raising up a static induction
charge, which, when discharged, produces a concentrated spark of no sensible duration, and therefore
much more luminous and audible than the former.
If one of the secondary terminals be connected with
the outside of a Leyden jar, and the other be continued until near the knob or wire connected with it,
a soft spark appears at such intervals for every successive current in the primary circuit.   This spark,
however, is *double*, for the electricity thrown into the
jar at the moment of induction is discharged back
again at the same place the instant the induction is
over ; the first discharge heats and prepares the air
there for the second discharge, and the two are so
nearly simultaneous as to produce the appearance of
a single spark to the unaided eye.—(*Faraday.*)
Ex. 12.—The difference in the thermal properties
of the induced current, with and without the intervention of the Leyden jar, is well shown by the
following excellent experiment, devised by Hearder :
—Connect a thermo-electrometer and a Lane's discharging electrometer with the terminals of the coil.
Upon an adjoining table place a disc of wood,
covered with tinfoil, exposing a flat surface of five or
six square feet, and connect it also with one terminal
of the coil.   Take a second similar disc of wood,
covered with tinfoil, and suspend it over the first one
by means of a string passing over pulleys, in a frame
so constructed as to admit of the second disc being
raised to the height of five or six feet above the
lower one.   Connect this disc by means of a flexible
wire, with the other terminal.   By this arrangement

the two terminals have virtually their conducting surfaces increased, and the sparks consequently are much brighter, though the thermo-electrometer is unaffected. If now the upper coated disc be gradually lowered, the sparks rapidly increase in power, and when they are within three or four inches of each other they assume the character of the discharges of a Leyden jar, and the thermometer begins to be affected. As the discs are gradually lowered still further, their faces being kept parallel to each other, the sparks become still louder, and the thermometer rises 15° or 20°, thus acting as coatings to a charged plate of air. On removing the upper plate these effects subside, and the spark reassumes its original character.

Ex. 13.—If a Leyden jar, coated with detached, diamond-shaped pieces of tinfoil inside and out, be connected with the terminals, it will be brilliantly illuminated during the whole time that the machine is in action. The best effects are obtained when the coatings are connected by two or three broad bands of tinfoil passing over the edge of the jar. If this be tolerably large, and if the rows of diamonds be so placed inside the jar that their horizontal points nearly touch one another at the centres of the circular holes cut in the diamonds of the outside, the discharge is exceedingly beautiful in a darkened room, far more so than with a common electrical machine.

Ex. 14.—If the discharges from a Leyden phial be made to pass over a lump of white sugar, or a crystal of alum, they will be beautifully illuminated; if through a fine iron two or three feet long, suspended by silk threads in a festoon, sparks, accompanied with brilliant scintillations, occur at every link. Should the chain be rusty, the brilliancy of the effect is increased.

Ex. 15.—Pass the discharge through a small heap

of gunpowder on the table of the universal discharger; it will be thrown about in all directions, but not ignited; now interpose a piece of wet string in the portion of this circuit between the discharger and the jar, the gunpowder will immediately be inflamed. This is precisely what occurs with ordinary electricity.

Ex. 16.—The following experiment is one of the most beautiful that can be made with the Induction Coil. It is called the "Cascade," and was thus described by Mr. Gassiot, who originated it:—Coat a beaker, about 4 inches deep by 2 inches wide, with tinfoil, so as to leave 1½ inches of the upper part naked. On the plate of an air-pump is placed a glass plate, and on it the beaker, covering the whole with an open-mouthed glass receiver, on which is placed a brass plate, having a thick wire passing through a collar of leathers; the portion of the wire within the receiver is covered with a glass tube; one end of the secondary coil is attached to this wire, and the other to the metallic plate of the pump. As the vacuum improves the effect is truly surprising: at first a faint, clear, blue light appears to proceed from the lower part of the beaker to the plate; this gradually becomes brighter, until by slow degrees it rises, increasing in brilliancy until it arrives at that part which is opposite, or in a line with, the inner coating, the whole being intensely illuminated; a discharge then commences from the inside of the beaker to the plate of the pump, in minute but diffused streams of blue light; continuing the exhaustion, at last a discharge takes place, in the form of an undivided continuous stream, overlapping the vessel, as if the electric fluid were itself a material body running over. If the position of the beaker be reversed, by placing the open part on the plate of the air-pump, and the upper wire either in contact with, or within an inch of, the outside of the vessel, streams of blue lambent flame appear to pour down the sides of the plate, while a continuous discharge takes place from the inside coating. On turning the commutator, so as to reverse the current, the cascade appears to flow upwards instead of downwards. This truly magnificent experiment has been arranged by Mr. Ladd so as to dispense with the trouble of exhausting the receiver on each occasion. Fig. 16 shows the apparatus. The cup, or beaker, is not

FIG. 16.

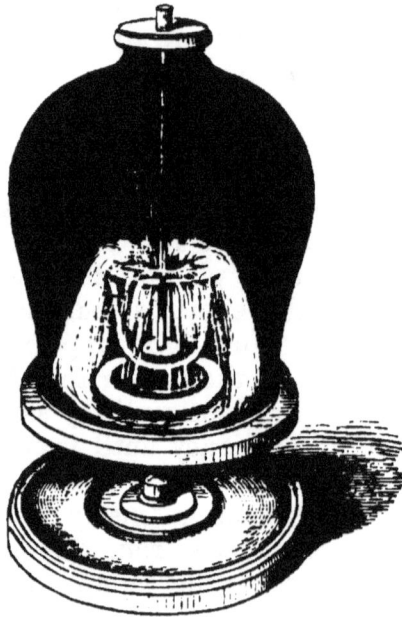

coated either inside or outside with tinfoil, but the wire through which the induced current is passed reaches to the bottom of the glass, and terminates in a metal disc. The receiver is permanently exhausted by an excellent pump, as shown in Fig. 17, it having been first, as well as the cap, firmly secured by cement. It is then screwed on to a stand, and thus rendered permanently portable.

It may here be remarked that no *ordinary* air pump is of much use in those experiments with the Induc-

Fig. 17.

tion Coil, in which a very good vacuum is required. In almost all the experiments about to be described, the most brilliant and gorgeous effects only appear when the exhaustion is approaching perfection. In the air-pump shown in Fig. 17, Mr. Ladd has arranged an extra barrel, a valve in which is mechanically opened by the movement of the piston; and though this is not adapted for the first process of exhaustion, it will, after the ordinary barrels have done their work, very materially increase the goodness of the vacuum.

But still more satisfactory results are obtained by the use of a modification of Geissler's mercury air-pump, shown in Fig. 18. It consists of two stout globular glass vessels, a tube from the bottom of the upper passing air-tight through the top of the lower and reaching nearly to the bottom. To the stem of the upper globe is sealed a lateral tube, and both it

FIG. 18.

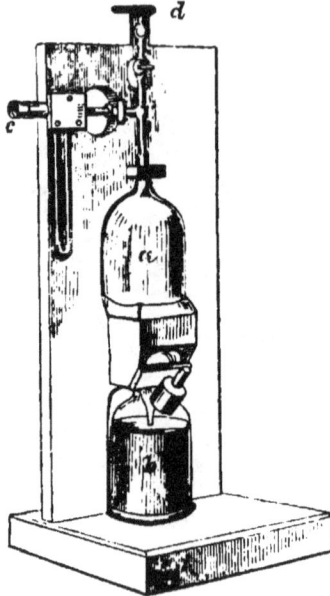

and the vertical stem from which it proceeds are furnished with stop-cocks. The cock on the lateral tube *c* is intended to communicate with the vessel to be exhausted, a syphon mercurial gauge being placed between them to register the degree of exhaustion. The stop-cock on the end of the vertical tube *d* is connected by means of elastic tubing with the air-pump or exhausting syringe. To the neck of the lower glass vessel *b* an elastic tube is also fitted, and this likewise is attached to a three-way cock on the air-pump.

The working of this apparatus is as follows :—The mercury (about 42 lbs.) being in the lower vessel b; the upper globe a is exhausted; the mercury rises and fills the vacant space. On opening the lateral stop-cock c, the air rushes in from the vessel to be exhausted, and the mercury sinks again partially into b. When the equilibrium is established, there will of course be mercury in both globes; the air in b is now exhausted, and the whole of the mercury is thus brought down to the lower vessel; the lateral stop-cock c is then closed; air is admitted into b, the mercury rises, and the air that has been drawn into a is exhausted at d. By continuing these operations, the air in the vessel to be exhausted becomes very highly attenuated.

Ex. 17.—De la Rive describes the following experiment in illustration of his theory of the Aurora Borealis:—Place the pole of a powerful electromagnet underneath the surface of mercury connected with the negative pole of a powerful voltaic battery; bring over and near it the positive pole armed with a charcoal point; a voltaic arc is formed, and the mercury is agitated above the magnet; luminous currents rotate round the pole, throwing out occasionally brilliant rays. This phenomenon of the rotation of electric light round a magnetic pole is exhibited in a most superb manner by the apparatus (Fig. 19). Into the brass cap of a large globular or egg-shaped glass receiver a soft iron bar, surrounded with a coil of covered copper wire, is fixed; the receiver is then exhausted. On sending the induced current through the vacuous receiver, a splendid band or riband of purple light makes its appearance, which immediately commences rotating round the iron rod, when that is converted into an electro-magnet by sending the current from a small voltaic battery through its surrounding coil; on turning the commutator, so as to change the direction of the induced current, the direction of the

rotation changes also.   In this truly magnificent experiment electric light takes the place of the con-

Fig. 19.

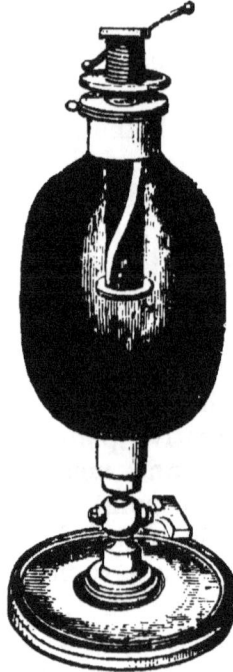

ducting wire in Faraday's discovery, mentioned in
page 4.   This experiment may be made more simple
with the little apparatus shown in Fig. 20, consisting of a small iron rod, enclosed air-tight in a
small receiver which is exhausted ; the iron rod
is surrounded with a glass tube, round which
there passes, towards the bottom, a metallic ring
attached to a wire which passes through the receiver ; a wire is also sealed into the top of the
glass, and through these wires the discharge is
made to pass ; the riband of purple light instantly

makes its appearance, and begins to rotate round the
iron rod, on placing the receiver on one of the poles

FIG. 20.

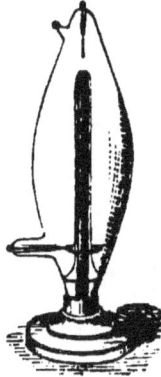

of a powerful permanent steel magnet or a small
electro-magnet.

Ex. 18.—Exhaust a tube, such as shown in Fig.
21, which may be from two to seven feet long, and
from $1\frac{1}{2}$ to 3 inches in diameter, having previously
connected the wires at each end with the terminals
of the coil. As the exhaustion proceeds, a splendid
Aurora Borealis fills the tube with corruscations, and
as the vacuum gets more perfect a broad crimson
riband is obtained, extending throughout the entire
length of the tube. Now turn the stop-cock very
gradually, so as to admit a very small quantity of
air, the effect of which is instantly seen by the disappearance of the riband and the re-appearance of
the corruscations; but these gradually die out as
the air enters. A few strokes of the pump, however,
bring them back again, and thus, by increasing or
diminishing the density of the air, the appearance
in the tube may be made to undergo corresponding
variations.

E

FIG. 21.

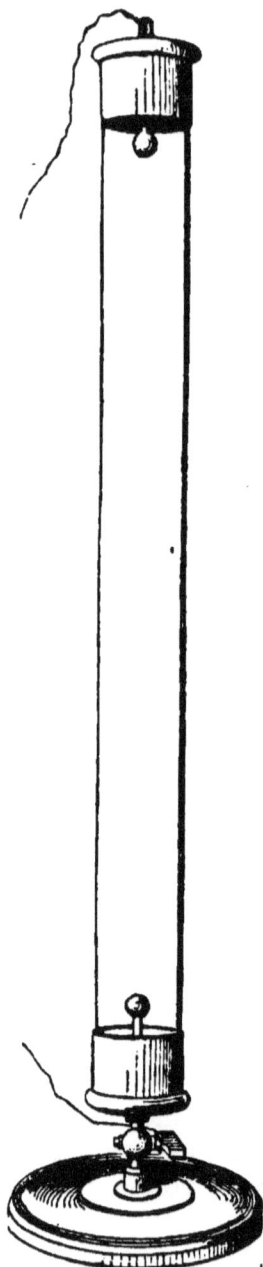

Stratifications in Electrical Discharges in Vacuo.—The striated condition of the electrical discharge in vacuo under certain conditions, was first announced by Mr. Grove, in a communication to the Royal Society, 7th January, 1852. The following was one of his first experiments : —A small piece of perfectly dry phosphorus was placed in a platinum capsule on the lower ball of the electric egg. To keep the receiver dry, a stick of caustic potash was suspended in a tube from the upper wire ; the exhaustion was then made as perfect as possible, when the crimson light became gradually furrowed with beautiful stratifications through a length which may be extended to 12 inches, and when once obtained, the experiments may be stopped, and after 20 minutes or more, resumed with more brilliancy than before. Mr. Grove afterwards found that the transverse dark bands could be produced in other gases when much attenuated, probably in all, and he thought the reason why they are more easily seen in phosphorus vapour is that, all the oxygen having been consumed, a better vacuum is formed. About the same time, Ruhmkorff noticed similar phenomena in an alcohol vacuum, and the subject engaged the attention of Massen, Quet, and Du Moncel.

The following modification of Grove's fine experiment with phosphorus is thus described by Mr. Jerry Barrett, and forms one of the most brilliant experiments that can be made with the Induction Coil:—"A glass tube not less than eighteen inches long by three inches in diameter is provided with a brass ball at the top attached to the ordinary flat brass plate and sliding wire, and at the bottom with a small metal cup half an inch in diameter attached to the nipple of the air-pump plate. This is to contain a piece of thoroughly dry phosphorus, about the size of half a pea; and the tube, after being rubbed inside with a warm cloth to insure the absence of moisture, is placed on the plate of the air-pump, and the top with the brass ball adjusted on it; after getting a good vacuum, the phosphorus will begin to glow, and contact with the coil should be made in the usual way—that is, the upper part of the tube should be connected by means of a fine wire with the arm of the instrument that is not provided with an ivory holder, and the other arm with the brass-work of the air-pump; it is very necessary in this as indeed in all experiments with the Induction Coil, that the connection should be carefully secured in order that no wire should become displaced when the room is darkened and so endanger the operator: contact being thus established, and the phosphorus allowed to glow for about five minutes, the commutator may be turned on, the phosphorus will then, by means of the electric spark, show signs of ignition, and the stream of electricity will become brilliantly stratified; then on continuing to work the extra barrel of the air-pump the light will become wider and fill the whole tube. Should the phos-

phorus at the commencement of the experiment have
shown sufficient activity, the stream of electricity
will now begin to assume a faint salmon colour, the
stratification becoming still more brilliant, until the
colour becomes white or silver, and the effect, to a
close observer, gorgeous in the extreme.

The changes of motion and form are produced by
means of the screw attached to the break, by reversing the commutator, and by varying the power
of the battery, but they are by no means certain.

It appears to be important for the success of this
beautiful experiment that the phosphorus, after a
good vacuum has been obtained, should be well
ignited by the electric current; but this does not
generally happen when the exhaustion has been
carried too far; it is equally necessary that the vapour
from the phosphorus be not too much deposited on
the surface of the glass tube so as to obstruct the
view, which would happen were the phosphorus too
soon ignited.

Sometimes the effect produced is that of a number
of cones of light chasing each other from below
upwards, and *vice versâ*; sometimes they are flat
tables of light, an inch or more apart; sometimes
they are rings apparently revolving or oscillating and
vanishing one into the other, and not unfrequently
the whole mass assumes the form of a cloud with no
motion whatever; sometimes there are two clouds,
and the effect of intercepting the current for a
minute or so is to bring back the stratification, which
lasts but for a very short time, and the cloud remains
as before, resisting all endeavours to produce stratification, except for two or three seconds after the
current is turned on. A very common effect is the
formation of one large column of little cones in
rapid motion, filling the whole tube, and reminding
one of the ripple of the sea by moonlight, and again

four or five streams of cones filling the tube from end to end all at the same time.

On more than one occasion, after varying the effects for upwards of an hour, I have succeeded in obtaining from sixteen to twenty layers of stratification, each layer being composed of two colours distinctly divided in the centre, the upper half green, the lower magenta, and *vice versâ* according to the directions of the current, exhibiting an effect similar to the very beautiful experiment in vacua produced by Mr. Gassiot with his large battery.

If, at the conclusion of these experiments, a small quantity of air be admitted into the tube, the effect will be extremely beautiful; it should be done as quickly as possible, and instantly checked; unless too much air has been admitted the stratification will not be destroyed, but a brilliant stream of magenta-coloured light will gradually blend with the whole: it is not always, however, that the original silver colour can be again restored.

These effects, which can be shown with still more magnificence in a larger tube, are continually varied, and this constitutes not the least of the charms of this remarkable experiment."

Ex. 19.—Fig. 22 is copied from the work of the last-named accomplished electrician, and very correctly represents the appearance presented in alcohol, wood-spirit, or turpentine vacua. When the poles are five or six inches apart, two distinct lights are produced, differing in colour, form and position. That round the negative ball and wire is blue—it envelopes it regularly; that round the positive is fire-red—it adheres to one side and stretches across towards the negative, and has for its lateral limits a surface of revolution about the axis of the receiver. On close examination this double light is seen to have a singular constitution; it is stratified, being composed of a series of brilliant bands, separated from each other by dark bands. In a good vacuum, the appearance is that of a pile of electric light. In

FIG. 22.

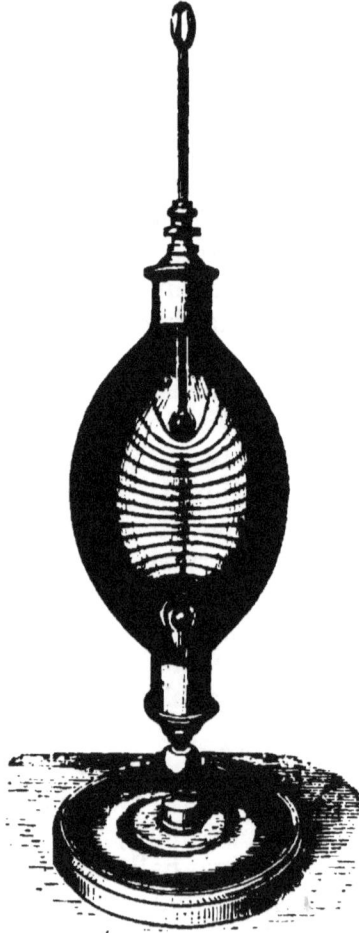

the red light, the brilliant bands approaching nearest to the negative ball have the form of capsules, the concave part being turned towards the ball; their position and figures are sensibly fixed, so that it is easy to see that there is a solution of continuity in passing from one to the other. The extreme capsule does not touch the violet light of the negative pole, being separated from it by a dark band, greater or less according to the nature and perfection of the vacuum, that with spirit of turpentine giving the greatest. It was found by M. Quet, that when a galvanometer was interposed at the circuit, no current was indicated as passing through the electric egg till the exhaustion was tolerably good, and the light continuous; the needle then became permanently deflected. A light, though less brilliant, may be obtained from one pole only, that of the exterior wire of the secondary, which possesses electricity of the highest tension; and if the vacuum be very good, this light may be made to bifurcate by placing the finger against the outside of the glass. If currents from two coils be made to circulate in opposite directions through the receiver, the red light disappears from the positive pole, giving place to a blue light—the positive and negative lights are now the same. The same occurs when a resistance is introduced into the induced circuit, as by interposing a condenser between one of the poles and one of the balls of the egg. A uniform blue light is thus obtained round both balls, which, with a good exhaustion, may be stratified.

Discharge in Torricellian Vacua.—The conditions necessary to enable the experimentalist to produce the phenomena of striæ or band discharge, have been stated to be:—1st. That the vacuum in the receiver should be as perfect as the air-pump can make it; 2nd. That care should be taken to absorb all trace of moisture; 3rd. That means should be used to introduce the vapour of naphtha or phosphorus, or other similar substances. In the barometrical vacuum, previous to the researches of Mr. Gassiot, detailed in the Bankerian Lecture (March 4, 1858), no striæ had been observed, the inductive spark being white and filling the whole tube ; by making these vacua, however, with great care, Mr. Gassiot has succeeded in obtaining stratifications very distinct and well defined.

Ex. 20.—Into the glass tube (Fig. 23) are sealed two platinum wires about eighteen inches apart ; the tube itself is twenty-eight inches long, and about five-eighths of an inch internal diameter; it is cemented into a brass plate, and when carefully filled with boiled mercury is placed on the open mouth of a receiver on the air-pump, the lower part of the tube being at the same time immersed in a basin of mercury ; by this arrangement the length of the discharge could be regulated from one-sixteenth of an inch to eighteen inches, either suddenly or very gradually, by allowing the air to enter into the receiver, or by exhausting it with the pump ; the vacuum is never perfect, a very minute bubble of air always remaining ; the stratifications are, however, very distinct when the discharge traverses the full length of eighteen inches. In this experiment a single cell of the battery may be used to excite the coil, and the condenser need not be used. If the discharge be made constantly in the same direction, the upper wire being negative, the upper portion of the tube, as far down as a line drawn even with the end of

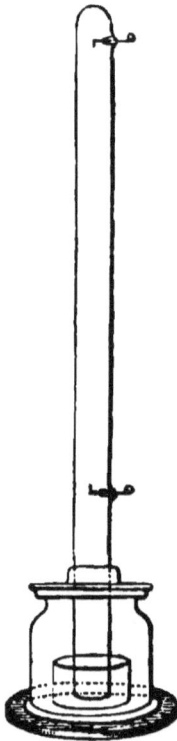

FIG. 23.

the wire, becomes covered with platinum in a minute
state of division; when this deposit is examined by
transmitted light it is transparent, presenting to the
eye an extremely thin bluish-black film; but by reflected light, either on the outside or inside, it has
the appearance of highly-polished silver, reflecting
the light as from the finest mirror. If the upper
wire be made positive, and the lower negative, as
soon as the mercury ascends above the negative wire
a beautiful lambent bluish-white vapour appears to
rise, while a deep red stratum becomes visible on the
surface of the mercury; as the mercury ascends in
the tube the stratified discharge from the positive
wire collapses, giving the appearance of a compressed
spiral; on exhausting, the mercury descends in the
tube, and the stratification expands as if the pressure
on a spiral spring had been removed. In the course
of his experiments on the inductive discharge through
Torricellian vacua, Mr. Gassiot found a great want
of uniformity in different tubes prepared in precisely
the same manner; in some, no stratifications at all
could be obtained, the discharge being clear, bright,
and white; in others, the discharge was a wavy line
unaccompanied with striæ; in others the stratification was confused and indistinct, while in others it
was clear and well-defined. He therefore prepared
some tubes by the non-boiling process, first proposed
for filling barometer tubes by the late Mr. Welsh, of
the Kew Observatory (for an account of which see
Phil. Trans., Vol. 146, p. 507), and with these he
obtained clear, well-defined, and distinct bands, not
only with the Induction Coil, but with the ordinary
electrical machine. The important feature in Mr.
Welsh's method of filling barometer tubes is the
perfect cleansing and drying of the tubes before the
introduction of the mercury, by sponging with whiting
and spirits of wine.

If the hammer of the contact-breaker be removed, and one of the terminals of the primary firmly fixed to a bright copper plate having a sharp edge, and the circuit completed by steadily pressing the end of the other wire on the plate, using one or two cells to excite the primary, no trace of any discharge will be perceived in the tubes; but if a sudden break of the battery circuit be effected, by bringing the wire quickly across the sharp edge of the plate, the stratifications immediately appear in the tube in a very distinct and beautiful manner; the more sudden the break, the more distinct will be the effects, and by using eight or ten cells they are distinctly visible on making contact. Contact with the battery may be also made and broken by dipping the wires in mercury. That the effects on making should not be equal to those on breaking contact, will be readily understood by considering that in the Induction Coil the inductive effects are principally due to the electro-magnetic condition of the iron core, and that while the iron wires require a certain time to reach their maximum power, they lose their magnetism *instanter* when contact is broken, provided the iron be very soft, and therefore the more suddenly the contact is broken, the more intense will the discharge appear in vacuo.

In experimenting with vacuous tubes, the operator should always pass the current in the same direction, as the emanation of the platinum particles, and the consequent deposit on the glass, only takes place from and around the negative wire, the positive end of the tube remaining clear and bright.

When the discharges of two separate coils were passed, by means of four platinum wires, through the same tubes, Mr. Gassiot found no signs of interference, the separate stratification of each coil remaining visible, although producing a degree of confusion from their interposition; he found also that

the stratifications were very powerfully affected by
the magnet, when the discharge is passing from wire
to wire; if a horse-shoe magnet be passed along the
tube, so as alternately to present the poles to different
contiguous positions of the discharge, the latter will
assume a serpentine form, in consequence of its
tendency to rotate round the poles in opposite directions, as the magnet in this position is moved up and
down the side of the tube.

Plücker, who has greatly distinguished himself by
his researches in various branches of physics, and particularly in electricity, has devised amongst many
others the two following experiments for illustrating
the action of magnetism on electrical discharge in
highly attenuated media:—In Fig. 24 two aluminum
rings are hermetically sealed into a glass tube, four

FIG. 24.

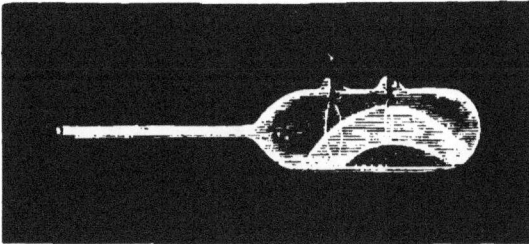

or five inches long and about one and a half inches
in diameter, the air in the tube is then exhausted as
perfectly as possible. On passing the discharge from
the Induction Coil between the two rings, the tube
becomes filled with a beautiful pale blue light. If
the small ring be made negative and the tube placed
between the poles of the electro-magnet (Fig. 3),
the moment the latter is excited the light arranges
itself in the form of a broad arc between the rings,
having a very beautiful appearance. On rendering
the electro-magnet passive the arc disappears, the

light in the tube re-assuming its diffused character;
but on re-exciting the magnet, the arc reappears. If
instead of two rings the terminals in the tube are
two aluminium wires, as shown in Fig. 25, the long

FIG 25.

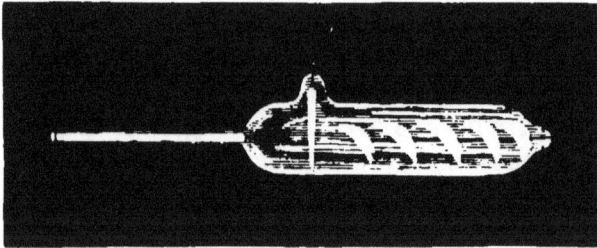

wire being made positive and the short wire negative,
the arc produced is very broad and brilliant.

Two Distinct Forms of Stratified Electrical Discharge.
—Ex. 21.—These are illustrated by employing the
simple tube shown in Fig. 26, which is thirty-eight

FIG. 26.

inches long, and is exhausted by Mr. Welsh's process; the wires, *a b*, are thirty-two inches apart; *CC'*
are moveable coatings of tinfoil, two inches long,
wrapped round the tube. When the discharges from
an Induction Coil are made from wire to wire, the
stratifications appear as already described; and if
the tube be placed in a horizontal position over the
pole of a magnet, the stratifications evince a tendency
to rotate as a whole in the direction of the well-
known law of magnetic rotation (2); but when the
discharge is made from coating to coating, or from
one wire to one coating, an entirely new phenomenon

arises, the stratifications have no longer a tendency to rotate as a whole, but are divided. If the tube be now placed between the poles of a powerful electro-magnet, one set of stratifications are repelled from, and the other attracted towards, or within, the bent portion of the magnet; when the tube is placed on the north pole the divided stratifications arrange themselves on each side of the tube, changing their respective positions when placed on the south pole, but in all cases each set of stratifications are concave in opposite directions. Mr. Gassiot, to whom this singular experiment is due, designates these discharges as the *direct* or conductive, and the reciprocating discharge. The former is that which is visible when taken from two wires hermetically sealed in a vacuum tube. This discharge has a tendency to rotate as a whole round the poles of a magnet; the latter is that which is visible in the same vacuum when taken from two metallic coatings attached to the outside of the tube, or from one coating and one wire. The induced charge is divisible by the magnet into two sets of stratifications, each set having a tendency to rotate round the pole of the magnet in opposite directions; the character of the electrical discharge, with relation to these two forms, can always be determined by the magnet.   ·

Discharge in Different Rarefied Media.—In dry hydrogen gas no discharge takes place from the Induction Coil, if the wires be separated in the tube beyond the striking distance in air; but when the gas is rarefied by the air-pump, the discharge first appears as a wavy line of bluish-grey colour; on continuing the exhaustion, and assisting the rarefaction by heating gently, the tube becomes filled with a luminous discharge to within about one inch of the negative wire; the stratifications appear gradually increasing in width as the vacuum becomes more perfect; and if care be taken to continue the pumping so as to prevent air being introduced, the tube can be sealed without the stratifications showing the slightest appearance of redness. If the extremity of a vacuum tube be presented to the prime conductor of an electrical machine, or to one of the terminals of an Induction Coil, a spark can be taken, and the glass will be perforated. The perforation is extremely minute, but sufficient, under the pressure on the vacuum, to admit air or gas; but, so slowly does the air or gas enter, that the experimentalist is enabled to note the gradual change which takes place during the progress of the discharges of the coil. Mr. Gassiot connected the extremity of a vacuum-tube, after perforation, by means of a tight-fitting gutta-percha tubing, to a glass cylinder containing fused chloride of calcium, through which air, hydrogen, oxygen, or nitrogen was permitted to pass into the vacuum. The result of many repeated experiments showed that with hydrogen and oxygen no change in the colour takes place ; with air or nitrogen the colour of the stratifications changes from bluish-grey to fawn, and ultimately to a deep red tinge ; and, during this addition of gas or air, the cloud-like stratifications gradually close, becoming narrower and narrower until they are utterly destroyed, passing to a mere luminosity filling the entire tube, and finally into the wave discharge.

The writer can confirm this description of the appearance presented when atmospheric air slowly makes its entrance into a vacuous tube. The experiment is interesting and instructive, although somewhat costly, and not one which amateurs will be very likely to repeat. On a late occasion, whilst exhibiting to an audience the beautiful stratifications in a carbonic acid vacuum, in a tube such as exhibited in Fig. 16, the striæ suddenly disappeared, and the discharge, which was at first nearly white, became first grey, then bluish-grey, and finally resolved itself into a riband of red light; this continued for some time, and then died away, and the discharge ceased to pass. On examining the tube, a very minute crack was observed proceeding from one of the platinum wires, probably the negative, which had become too highly heated. Too much battery power had been employed. The accident is related as a warning to those inexperienced in those experiments to be very careful not to excite the coil too strongly whilst passing the inductive discharge through the vacuous tubes. Two cells of Grove will be found amply sufficient, and even with this power it will be well to relax somewhat the spring of the contact-breaker.

Influence of Temperature. — The following results were obtained by Professor Faraday and Mr. Gassiot, in Torricellian vacua, through a range of upwards of 700 deg. Fahr. A vacuum which gave good cloud-like stratifications, exhibited no change when the temperature was lowered to $+ 32°$; but at a temperature of $- 102°$, obtained in a bath of ether and solid carbonic acid, all traces of stratifications were destroyed, and in this state the red or heated appearance of the negative wire disappeared, the discharge filling the entire vacuum with a white luminous glow. On the temperature being raised by the application of heat to the mercury, the stratifications re-appeared. When the mercury was boiled, indicating heat of upwards of $+ 600°$ Fahr., the stratifications were all destroyed, but in this case the discharge passed along the mercury as it condensed in the cooler part of the tube. When the mercury was frozen the stratifications disappeared, and the discharge did not illuminate the entire length of the tube, but merely the terminals. In this state, when a horse-shoe magnet was brought near the tube, the cloud-like stratifications immediately appeared from the positive wire, very distinct and large, but not so clearly separated as when the tube was at its normal temperature.

Discharge in Carbonic Acid Vacua.—At the suggestion and with the assistance of Dr. Frankland, Mr. Gassiot prepared tubes in which the carbonic acid with which they were filled was absorbed, after exhaustion, by a good air-pump, by caustic potash. In vacua obtained by this process, the discharge from an Induction Coil is first in a white wave line, strongly affected by the magnet, or by the hand when placed on the tube. In this state the discharge does not generally present the stratified appearance, or if present the stratifications are only near the positive terminal. After a time, however, as the carbonic acid becomes absorbed by the potash, the stratifications gradually appear more clearly defined; they assume a conical form, and, lastly, the cloud-like appearance of the best mercurial vacua. After this, under some conditions, the stratified appearance entirely ceases, the whole length of the tube being filled with faint luminosity. When in this state, if the outside of the tube be touched with the finger, pungent electrical discharges arise, and sparks one-eighth of an inch in length can be elicited. The appearance presented when the discharge was sent through a tube four inches long, the wires which

FIG. 27.

were one inch apart being terminated with gas-coke balls one-eighth of an inch in diameter, was as shown in Figs. 27 and 28. On the positive coke, minute luminous spots were visible ; the negative coke was surrounded with a brilliant glow. At intervals,

FIG. 28.

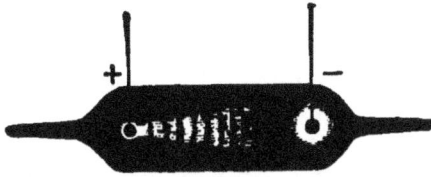

apparently by some energetic action, flashes of bright stratified light would dart through the vacuum, but by carefully adjusting the contact-breaker, the discharge could be made to pass, without to the eye affording any appearance of an intermittent discharge.

A large egg-shaped glass vessel, the globular portion being eighteen inches long and seven inches in diameter, was made under Mr. Gassiot's direction ; the wires were twenty-two inches apart, and caustic potash was placed in the narrow end. It was filled with carbonic acid, and exhausted by Dr. Frankland's process. A portion of the potash being heated by a spirit-lamp, in about two months the discharge assumed, in a very marked manner, the character of large distinct clouds, most clearly and separately defined ; they were strongly affected by induction as the hand approached the globe, presenting a very striking appearance. There was a slight tinge of red, showing that a very minute quantity of air remained ; the cloud-like stratifications extended to the entire diameter of the vessel.

Fig. 29 is another form of apparatus. The tube is

FIG. 29.

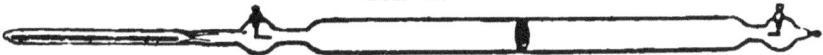

fourteen inches long and about one inch internal diameter ; it has a glass division in the centre,

F

perforated with a hole about one-eighth of an inch in diameter. The striæ on the negative side are here very clearly defined, while on the positive side they are indistinct. When the discharge has assumed the cloud-like appearance, the aperture in the diaphragm only affects the discharge by contracting the cloud which passes immediately through it. That the passage of the discharge depends upon the presence of matter, and the stratifications probably to pulsation, or impulses of a force on highly attenuated matter, seem to be rendered probable by the fact that, in some of Mr. Gassiot's tubes, in which the absorption of the carbonic acid by caustic potash contained in one end of the tube was complete, no discharge could be made to pass; the same was the case with other tubes containing, besides caustic potash, fused chloride of calcium, sulphur, and selenium.

Luminous Discharge of Voltaic Batteries when examined in Carbonic Acid Vacua.—When the discharge from a water-battery of 3,520 cells was sent through carbonic acid vacua tubes, stratified discharges, similar in character to those of the Inductive Coil, were obtained, and Mr. Gassiot found that whenever the potash in any of the tubes was heated the discharge entirely ceased. From the steady deflection of the galvanometer needle placed in the circuit, the discharge had the appearance of being continuous, but closer examination showed them to be intermittent. From a Daniell's battery, consisting of 512 series of elements, no stratified discharge could be obtained through any of the vacuous tubes, but in one, a brilliant glow was observed round the negative, and a trifling luminosity round the positive coke ball terminal. With 400 series of Grove's nitric acid battery, each cell carefully insulated, the most magnificent effects were obtained. The different vacuum tubes used were introduced between one of the copper discs of a micrometer-electrometer and the battery, as also a galvanometer. By this arrangement the circuit could be gradually completed without any risk of disarranging the apparatus, and the spark discharge obtained before the copper discs of the micrometer-electrometer came into contact. Dr. Robinson thus describes an experiment which he witnessed in Mr. Gassiot's laboratory, with a tube 24 inches long and 18 in circumference, one terminal being a copper disc, 4 inches in diameter, and the other a brass wire :—" On the completion of the current, the discharge of the battery passed with a display of magnificent strata of most dazzling brightness. On separating the discs, by means of a micrometer screw, the luminous discharges presented the same appearance as when taken from an Induction Coil, but brighter. On the copper plate in the vessel there was a white layer, and then a dark space about an inch broad ; then a bluish atmosphere, curved like the plate, evidently three negative envelopes on a great scale ; when the plate was positive the effect was comparatively feeble." Between coke terminals a stream of light, of intolerable brightness, was presented, in which strata could be observed through a plate of green glass ; this soon changed to a sphere of light on the positive ball, which became red hot. On heating the caustic potash, the discharge again burst into a sun-like flame, subsequently subsiding in three or four large strata, of a cloud-like shape, but intensely bright. The appearances presented when the potash was heated are depicted in Figs. 30 and 31. Mr. Gassiot arranged the apparatus by attaching gold-leaf electroscopes to both terminals, and introducing the galvanometer so as to enable him to examine more carefully the action that would take place when the potassa was heated. On heating the potassa, the fine negative glow was developed; the leaves of the electroscope did not close, but as the

FIG. 30.

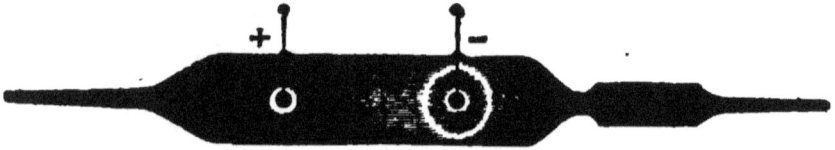

FIG. 31.

negative glow increased, the needle of the galvanometer was suddenly deflected, immediately returning to zero. As more heat was applied, a small globe of light appeared on the positive ball, and the needle was gradually deflected 40° to 50°. On withdrawing the lamp, as the potash cooled, the positive glow disappeared, the needle of the galvanometer receded, the glow on the negative remaining more or less brilliant; this action and reaction alternating as the heat of the lamp was applied to, or withdrawn from, the potash. When the heating of the potash was further increased, four or five cloud-like and remarkably clear strata came out from the positive ball, (Figs. 32 and 33), and these were quickly followed by a sudden discharge of the most dazzling brightness, which remained for several seconds. The needle of the galvanometer was suddenly and violently deflected. By these and many other equally striking experiments, Mr. Gassiot proved that the luminous and stratified appearances obtained in carbonic acid

vacua do not arise from any peculiar action of the
Inductive Coil, whatever the real cause of the
phenomena may ultimately prove to be.

FIG. 32.

FIG. 33.

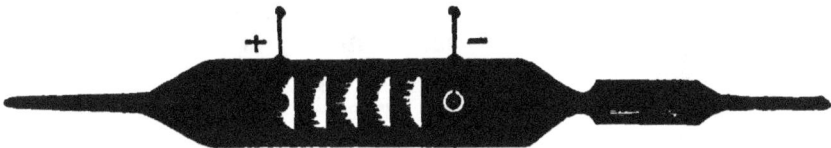

More recently (Proceedings of the Royal Society,
Dec. 11th, 1862), Mr. Gassiot has studied minutely
the stratified appearance in electrical discharges,
employing in his experiments a battery of 3,360 pairs
of elements charged with salt and water and carefully insulated, and an extensive series of the sulphate
of mercury battery also carefully insulated. "The
discharge of the battery," he observes, "is much
more sensible to the slightest variation of the state
of tension in the vacuum tubes, than that of the
Induction Coil; the sudden disruption in the discharge of the latter presenting greater obstacles to
the more careful study of the phenomena than is
offered by the direct discharge of the battery."

The battery was arranged in three groups, each
consisting of 1,520 pairs of elements, and the general
practice was to place the experimental tube between
either the one or the other of them—the positive
terminal of one battery A being attached to one wire,

and the negative terminal of another battery в to the other wire, the opposite poles of а and в were then connected and the circuit thus completed. In order to vary the resistance at pleasure, two tubes containing distilled water were included in the circuit. By varying the depth to which the wires attached to the terminals of the battery were plunged in one or both of the tubes the resistance could be regulated with great precision. When the discharge was passed through a tube 20 inches long and 4 inches in diameter, one terminal consisting of aluminum, cup-shaped, about 3 inches in diameter, and the other a wire of the same metal, it was, when examined by a revolving mirror, intermittent, and distinct sounds were heard when a magnet was presented to the tube. With a tube about 5 inches long, the terminals being balls of aluminum, the discharge from the entire battery was of dazzling brilliancy, exhibiting twelve or fourteen striæ. When the water resistance tubes were interposed, the phenomena were in the highest extraordinary. The tubes were each 18 inches long, and were connected with each other from the bottom by a wire. As soon as the battery wires touch the surfaces of the water, a faint luminous discharge is observed at each ball of the vacuum tube ; as one wire attached to the negative is slowly depressed, the two luminous discharges appear to travel towards or to attract each other. Depressing the wire very gradually, the positive discharge becomes sharply defined, the negative retaining much of its irregular termination, but each separated from the other by a dark interval of about 1 inch in length. As the wire is further depressed in the water, the brilliancy of the positive and negative glows increase ; and when about 3 inches of one wire have been immersed in the water a single clearly defined luminous disc bursts forth from the positive, remaining steady and apparently fixed. As the wire is further depressed in the water, the luminous discharge at the positive slowly progresses along the tube until another bright disc appears remaining stationary like the first. When the resistance is again reduced by depressing the wire still further into the water a *third* luminous disc is developed, and at 18 inches depression or the entire length of one column of water a *fourth* disc is observed. In this state, while the four luminous discs are stationary, if the wire attached to the positive terminal of the battery is depressed, the luminous discs gradually closing on each other become more compressed, when a *fifth* is developed. By continuing in this manner gradually to diminish the resistance new discs start one by one into view, until the number is increased to thirteen or fourteen. On gradually raising the wires, the discs one by one disappear. Many other beautiful phenomena are described by Mr. Gassiot in this memoir, to which we must refer our readers. In summing up his results he says :—" May not the dark bands be the *nodes of undulations* arising from impulses proceeding from positive and negative discharges ? or can the luminous stratifications which we obtain in a closed circuit of the secondary coil of an induction apparatus, and in the circuit of the voltaic battery, be the representation of *pulsations* which pass along the wire of the former and through the battery of the latter, impulses possibly generated by the action of the discharge along the wires ?"

Geissler's Vacua Tubes.—It would be quite impossible, by any description, to do justice to the extreme beauty of the phenomena observed when the inductive discharge is passed through many of the tubes so ingeniously prepared by M. Geissler, of Bonn ; neither indeed could any description, however correct, serve any useful purpose, as, in conse-

quence of the almost impossibility of preparing two
tubes precisely alike as to form, and as to the exact
condition of the attenuated media they enclose, it is
very difficult to find two which present the same
appearances; moreover, we are for the most part
ignorant of the nature of the matter with which
these tubes have been filled, so that Mr. Gassiot, in
his investigations, was compelled to prepare his own
tubes. "Though," he writes, "I had the opportunity
of experimenting with upwards of sixty of Geissler's
vacua tubes, in which many beautiful and novel
results are produced, not being able to ascertain with
accuracy what is the gas, which, however attenuated,
must remain in each tube, and from most of them
being constructed of a varied form in consequence of

Fig. 34.

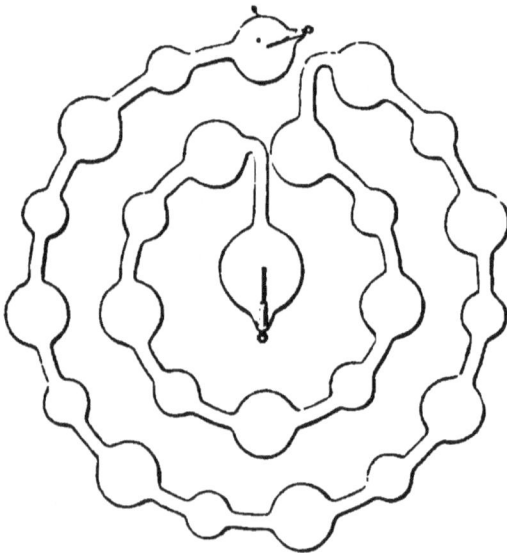

which the discharge presents, in the several portions
of the same tube, an entirely different appearance,
both of colour and in form of stratification, I was

reluctantly compelled to lay them aside, and either to charge and exhaust each tube myself, or have them charged and exhausted in my presence." By way of reference, we have, however, figured some of the most striking of these vacua tubes, and would beg to remark that Mr. Ladd, from whom they may all be procured, is constantly receiving novel accessions from Germany. In the frontispiece the inductive discharge is represented as passing through a spiral tube, in which twenty-five bulbs have been blown. In the tube in the writer's possession, the light is white, with a pale green tinge, the effect of which is greatly exalted by placing behind it a black curtain; after the discharge has ceased the tube remains for some seconds phosphorescent. A similar tube is shown in Fig. 34. The operator is warned in this and all the vacua experiments, not to employ more than two, or at most three, battery cells.

Fig. 35 is a similar spiral tube, containing only sixteen bulbs.

FIG. 35.

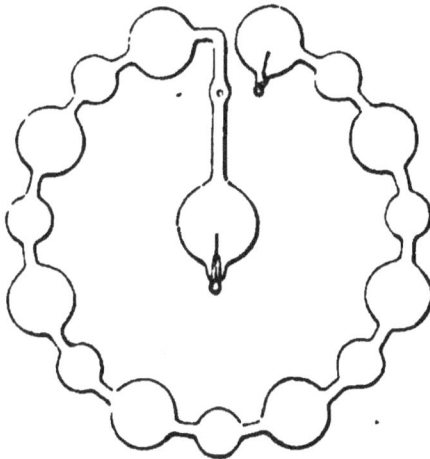

Fig. 36 represents a very brilliant design adapted

for an illumination   The thick parts of the letters
contain the fluorescent solutions (p. 92), producing
a very pleasing contrast.

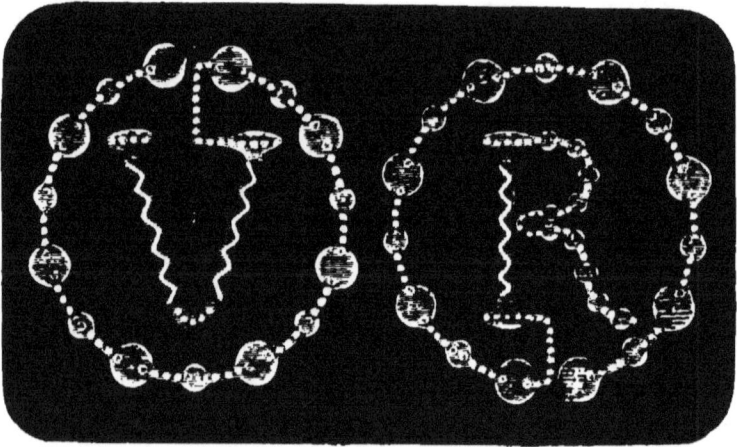

Fig. 37.—Bulbs pale green, tho connecting tubes
pale red, with stratifications; phosphorescent after
the discharge has ceased.

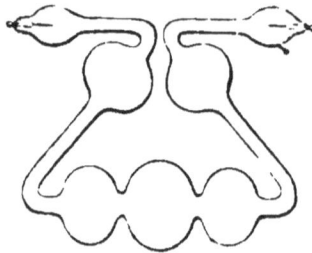

Fig. 38.—Bulbs pale green, connecting tubes red,
with stratifications; phosphorescent.

Fig. 39.

Fig. 39 and 40.—These are beautiful experiments; the bulbs in Fig. 39 vary from 5 inches to 3 in diameter; the centre bulb, and the two smaller ones, are filled with a pale green light, with magnificent stratifications, the connecting tubes pale red; the bulb in connection with the negative terminal is of a

Fig. 39

Fig. 40.

delicate mauve colour; that connected with the positive is red; but by turning the commutator, these colours change places; the stratifications are urged from the negative towards the positive terminal. In Fig. 41 the left-hand bulb is of uranium glass,

Fig. 41.

which gives the characteristic yellow colour; the tube connecting it with the centre bulb is of lead glass, and the colour of the light is blue, and the centre bulb is pale green.

Fig. 42.—The bulbs are filled with a pale green light, except the terminal ones; the negative being

Fig. 42.

mauve and the positive pale red; the connecting tubes are pale red and beautifully stratified; when the discharge is suspended the bulbs remain for some seconds phosphorescent.

Fig. 43.—The spirals are made of uranium glass,

Fig. 43.

which come out of a fine green colour; the centre bulb is pale red; the positive bulb is red and the negative mauve; this is a nitrogen vacuum.

Fig. 44.—The spirals in this tube are enclosed in two separate tubes, one of which is filled with solu-

Fig. 44.

tion of disulphate of quinine and the other with water, through which a few drops of infusion of horse-chesnut bark have been diffused; the light on the spiral tube is red, surrounded on the quinine side by a beautiful blue, and on the œsculine side by a fine green. Other fluorescent liquids with which tubes of various shapes are filled are:— Solution of *amido-phthalic acid* which gives a fine *uranium* colour; solution of *amido-tcrephalic acid* which gives a magnificent blue; and *pavine* from willow-bark which gives a rich *brown*. Tincture of the seeds of *stramonium*, of *turmeric*, and of *chlorophyll*, likewise exhibit the phenomena of fluorescence.

There are certain solids which, after being exposed to solar light or to the sparks from the Induction Coil, continue to emit light for a considerable time. Among these substances *Cantor's phosphorus* (sulphide of calcium), and *Baldwin's phosphorus* (fused nitrate of calcium), and *Bologna stone* (sulphide of barium), have long been known; but the researches of Becquerel have shown that this property of *phosphorescence* is possessed by many other compounds, amongst which may be mentioned:—Sulphide of strontium, and the salts of the alkalies and alkaline earth generally; certain varieties of diamond and of fluor spar, and most transparent objects, particularly those of an organic origin. Small boxes are fitted up with tubes filled with various phosphorescent solids, and the phenomena produced by exposing these tubes for a few seconds to the intense light of the induction spark and then removing them into the dark are very striking.

Fig. 45.—When the vacuum in this tube is pure

FIG. 45.

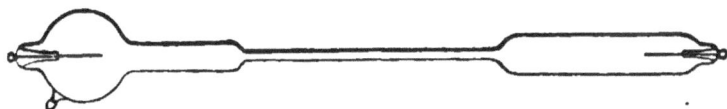

carbonic acid, the light is white; when pure hydrogen, the centre bulb is pale green, stratified, and the connecting tubes vivid red; when from bi-chloride of tin, the bulbs are pale blue, and the connecting tubes yellow.

Fig. 46.—This form of tube is intended for medical use; it furnishes the practitioner with an excellent

FIG. 46.

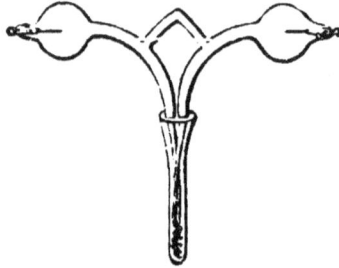

and convenient means of examining the throat, for which purpose the tube enclosing the spiral is introduced at the mouth, and the inductive discharge passed through the bulbs, which have been filled with carbonic acid and well exhausted; a brilliant white light is produced, which illuminates the interior of the mouth and throat.

Tubes of the simple shape shown in Fig. 47, are

FIG. 47.

well adapted for observing the nature of the discharge, and the stratification in different gaseous vacua.

When vacua tubes are so arranged that continuous rotatory motion can be given to them, very beautiful effects are produced. Fig. 48 shows the mode in which Mr. Ladd mounts his tubes; they may, of course, be varied *ad libitum* in shape and in mode of preparation. By turning the wheel with different degrees of velocity, and altering the frequency of the discharge and direction of the current, various optical phenomena are brought out, which greatly

FIG. 48.

increase the magnificence of the display. This apparatus is called Gassiot's revolving star.

The following is a summary of the effects produced by the electric discharge through Torricellian vacua. (*Grove*).

If the vacuum be equal to that generally obtained by an ordinary air-pump, no stratifications are perceptible; a diffused lambent light fills the tube. In a tube in which the rarefaction is carried a step further, narrow striæ are perceptible, like those obtained with phosphorus vapour. A step further in rarefaction increases the breadth of the bands. Next we get the conical or cup-shaped form; and then, the rarefaction being still higher, we get a series of luminous cylinders of an inch or so in depth, with narrow divisions between them. Lastly, with the best vacua which have been obtained, there is neither discharge, light, nor conduction. The fact of non-conduction by a very good Torricellian vacuum was first noticed by Walsh, subsequently carefully experimented on by Morgan, and afterwards by Davy; the latter did not, however, obtain an entire non-conduction, but a considerable diminution both of light and conducting power. From these experiments it may be concluded that in vacuo, or in media rarefied beyond a certain point, electricity will not be conducted, or, more correctly speaking, transmitted, an extremely important result in its bearing on the theory of electricity.

The following is Mr. Grove's view of the *rationale* of the phenomena of stratification. When the battery contact is broken, there is generated the well-known induced current in the secondary wire, in the same direction as the original battery current, to which secondary current the brilliant effects of the coil are due; but in addition to this current in the secondary wire, there is also a secondary current in the primary wire, flowing in the same direction, the inductive spark at the moment following the disruption of contact, completing the circuit of the primary, and thus allowing the secondary current to pass. This secondary current in the primary wire produces, in its turn, another secondary, or what may be termed a tertiary current in the secondary wire, in an opposite direction to the secondary current. There are thus almost synchronously two currents in opposite directions in the secondary wire; these, by causing a conflict or irregular action on the rarefied medium would give rise to waves or pulsations, and might well account for the stratified appearance. Mr. Grove quotes the following experiment as strongly in favour of this theory. It is obvious that the secondary must be more powerful than the tertiary current. Now, supposing an obstacle or resistance placed in the secondary circuit which the secondary current can overcome, but the tertiary cannot, we ought, by theory, ·to get no striæ. If an interruption be made in the secondary current, in addition to that formed by the rarefied medium, and this interruption be made of the full extent which the spark will pass, there are, as a general rule, no striæ in the rarefied media, while the same vacuum tube shows the striæ well if there be no such break or interruption. The experiment was shown by Mr. Grove, in a lecture, at the Royal Institution (January 28, 1859), with a large vacuum cylinder (16 inches by 4) and Mr. Gassiot's micrometer-electrometer; this tube showed numerous broad and perfectly distinct bands when the points of the micrometer were in contact; but when they were separated, to the fullest extent that would allow sparks to pass, not the slightest symptoms of bands or striæ were perceptible, the whole cylinder being filled with an uniform lambent flame. With a spark from the prime conductor of the electrical machine, the striæ do not appear in tubes which show them well with coil; occasionally, and .

in rare instances, striæ may be seen with sparks from
the electrical machine, but not when the spark is un-
questionably single. All this Mr. Grove thinks is in
favour of his theory; but without regarding that as
conclusive, or as a proved *rationale*, it is clearly de-
monstrated by the above experiments, that the
identical vacuum tubes which show the striæ, with
certain modes of producing the discharge, do not
show them with other modes, and that therefore the
striæ are not a necessary condition of the discharge
itself in highly-attenuated media, but depend upon
the mode of its production. Certain experiments,
described by Mr. Gassiot (*Phil. Trans.*, 1859), do not
harmonize with Mr. Grove's view. He found that
when a Leyden discharge was sent through a vacuum
tube, stratifications, as clear and as distinct as those
from an Induction Coil, may be obtained by reducing
the intensity of the discharge, by the introduction
into the circuit of a piece of wet string; he hence
inferred that in Mr. Grove's experiment, the absence
of striæ, when the circuit was interrupted, was due
to the heightened intensity of the discharge. He
repeated Mr. Grove's experiment with the large
cylinder, and obtained a similar result; the stratifications were entirely destroyed when the secondary
circuit was interrupted, but they were restored when
a second interruption was made in the circuit, and
this closed by a wet string; in this case it is evident
that the appearance of the striæ does not depend
upon the conflict of secondary and tertiary currents,
but upon the manner in which the discharge passes.
Mr. Gassiot found, moreover, that when, by means
of an interrupted discharge, the stratifications are
destroyed, they are reproduced in a carbonic acid
vacuum tube when heat is applied to the caustic
potash; here the increased resistance arises from the
greater density of the matter formed in the tube;

and the experiment favours the view of Mr. Gassiot,
viz., that the stratifications arise from the effect due
to pulsation or impulses of a force acting on highly
attenuated matter.

*Spectra in highly rarefied Gases of different kinds,
during the passage of the Electrical Discharge.* — In
order to observe and analyze the spectra, Professor
Plücker concentrated the luminous electrical discharge current in thermometer tubes whose internal
diameters were nearly the same for the different
gases examined, being about 0·6 millimetre. Fig.
50 shows the form of the perfect separate gas tubes,
as well as the manner in which they may be connected on a piece of board, so that the narrow parts

Fig. 50.

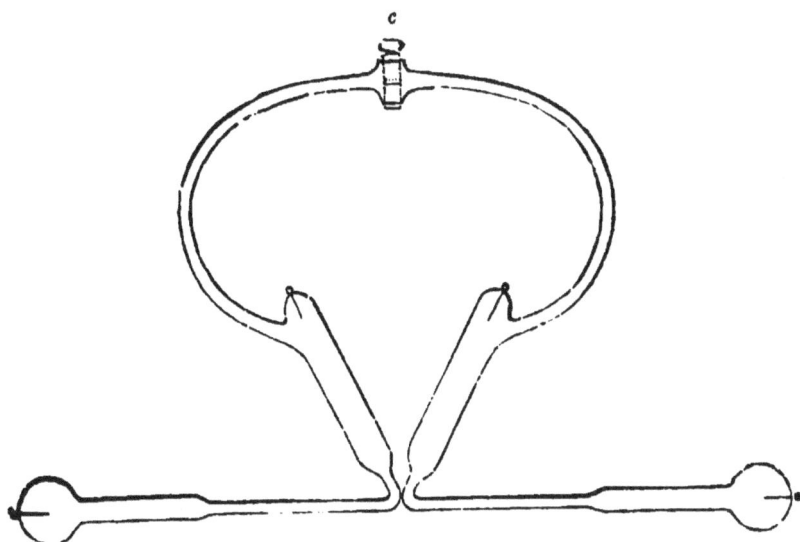

of both (at the parts where they are bent at an angle
of rather more than 90°) touch one another, and
have exactly the same direction. By turning the
glass cock (*c*) the gases in the two tubes could be
put into communication. The spectra were observed

by means of a telescope (such as that employed by Fraunhofer, in the observation of the lines of the solar spectrum) without angular measurements. This was set up at a distance of from 4 to 5 metres from the vertical line of light in the tube. The flint-glass prism, whose refractive angle was 45 degrees, was fastened immediately before the object glass, whose aperture was 15 Paris lines.

1. *Hydrogen.*—Almost the whole of the light is concentrated into three bands, — namely, a dazzling red, at the extremity of the spectrum; a beautiful greenish blue; and finally a violet of inferior brightness, whose distance from the greenish blue is about two-thirds of the distance of the latter from the dazzling red. In the narrow tube the electric light stream appears red.

2. *Nitrogen.*—In the spectrum of this gas all the colours are fine, none of them being faded, as in the broad spaces lying between the bright bands of the hydrogen spectrum. In the spaces of the red, orange, and yellow, there are about fifteen narrow dark-grey lines at nearly equal distances apart; six of these belong to the orange and yellow; both of these colours are beautiful. The red, in the direction away from the orange, is shaded off into brown, but becomes brighter and purer towards the extremity of the spectrum, which stretches beyond the dazzling red bands of the hydrogen spectrum. A broad green space is separated from the yellow by a narrow black band. The greater part of this space appears shaded with black in the direction away from the black band. On a more careful examination, this shading is seen to consist of very fine black lines, which are at equal distances apart, but nearer together than the previously mentioned bands on the red, orange, and yellow. The rest of the green space is again subdivided. The green is bordered by

two beautiful bright blue bands, which are sharply separated from one another, and from the green, by narrow black bands. The blue and red violet ends of the spectrum form nine sharply-bordered violet bands, alternating with dark ones. The fourth and fifth bright bands, separated by a black band, possess the most light; the four following ones are less prominent; the last one, however, which forms a sharp boundary to the whole spectrum, is the most distinct. The light of the discharged current in the narrow tube is *yellowish-red*.

3. *Carbonic Acid.* — Six bright bands sharply separate the bright portion into five spaces, of which the two first are of equal breadth; the third, and especially the two last, are somewhat broader. The first of the six bands is situated on the extreme boundary of the red, the second is reddish-orange, the third greenish-yellow, the fourth green, the fifth blue, and the last violet. Both of the two first spaces are divided into three equally broad subdivisions by narrow black-grey bands, of which two always border upon the bright band. The first space is brown-red; the second dirty-orange and yellow; the third and fourth spaces are of rather faded green, and much subdivided by different degrees of shading; the fifth space, which is very faded, is divided into two equal spaces, which are shaded off from the red side towards the violet. After the last-mentioned violet band, another dark portion of the spectrum occurs, about as wide as the red-yellow portion. In this dark portion, three spaces are separated by three prominent and well-marked violet bands, whose breadth is of the same value as that of the before-mentioned six bands. The last of the three violet bands forms the visible boundary of the spectrum. The first of these three spaces, which is contiguous with the above six bright bands, is somewhat broader

than the third. Both are perfectly black. The second and middle space is about as broad as the first and third together, and is of a very dark violet colour. The first band, which at the moment of commencing was of an especially brilliant red, lost almost the whole of its brightness after the streams had passed through the tube for a long time. This was occasioned by the decomposition of the gas into carbonic oxide and oxygen, the latter combining with the platinum of the negative electrode, and forming oxide of platinum, which was deposited of a yellow colour upon the neighbouring internal glass surface.

4. *Ammonia.*—On exhausting a single tube that has been filled with ammonia, and passing the current, a spectrum was produced which was evidently the result of the superposition of the two spectra for hydrogen and nitrogen ; the ammoniacal gas was immediately decomposed into its constituents, and it was not possible to obtain the spectrum of the chemically combined gases. When one of the double tubes, Fig. 50, was filled with carbonic acid, and the other with hydrogen, and then exhausted as far as possible, a greenish white light was obtained in one tube and a red light in the other. On now putting the two gases into communication, by opening the stop-cock, *c*, and observing the spectrum of the carbonic acid through the telescope by the prism, a dazzling red line was at first seen merely flickering now and then at the boundary of the spectrum, and soon took up and maintained a constant position ; this was the red band of the hydrogen gas. The colour of the light in the two narrow tubes was the same—the two spectra had become constant and identical in kind.

5. *Oxygen.*—A good spectrum could not be obtained with this gas, on account of its gradual disappearance and combination with the platinum of the negative

electrode. Oxide of platinum, of a yellow colour, was deposited upon the neighbouring internal glass surface, showing, by reflected light, the colours of Newton's rings in a very beautiful manner. If the tube contains traces of hydrogen or nitrogen, metallic platinum is transferred to the glass surface. The colour of the electric light current in the narrow tube was at first red; it passed through a flesh colour to a green, and then through blue to a reddish-violet, and then became extinct, proving that no current can exist in absolute vacuæ.

6. *Binoxide of Nitrogen.*—This gas was decomposed, the spectrum for nitrogen being obtained with a modification evidently attributable to pure oxygen (a bright band near the red); this was gradually extinguished, and the result was the formation of the pure spectrum of nitrogen gas of a splendour which Plücker had never before witnessed. Binoxide of nitrogen, present in so small a quantity as to be scarcely recognizable by the most sensitive balance, was thus chemically analyzed; with nitrous acid the red band due to the oxygen was at first of great brilliancy, but it gradually disappeared; the same was the case with protoxide of nitrogen.

7. *Aqueous Vapour.*—The electric current in the narrow tube showed the most beautiful deep red.

The spectrum was that of pure hydrogen, with its three prominent bands, in comparison with the brightness of which the rest of the luminous divisions were so insignificant that here the shading off of colour and luminous intensity was scarcely to be recognized. The aqueous vapour had separated into its simple constituents.

8. *Iodine, Bromine, and Chlorine.* — Pure spectra have not yet been obtained with these substances, because the manner in which the tubes have hitherto been made did not admit of complete exclusion of the

air. That which the three spectra have in common, and by which they are distinguished, as far as present observations extend, from all other gas spectra, consists in lines of light, which at first are constant, but afterwards only flickering, and whose width is about the same as that of the narrow Fraunhofer's black lines. The chemical results hitherto obtained are summarized briefly by Plücker, as follows :—

1. Certain gases (oxygen, chlorine, bromine, and iodine vapour) combine more or less slowly with the platinum of the negative electrode, and the resulting compounds are deposited upon the neighbouring glass surface. When the gases are pure, we thereby gradually approach to a perfect vacuum.

2. Gases, which are composed of two simple kinds (aqueous vapour, ammonia, nitrous oxide, nitric oxide, nitrous acid), immediately split up into their simple constituents, and then remain unchanged, if these latter (ammonia) do not combine with the platinum. If one of the constituents is oxygen (in water, and the different stages of oxidation of nitrogen) this gradually disappears, and the other gas alone remains.

3. If the gases are composed of oxygen and a solid simple substance, complete decomposition by the current only takes place gradually, while the oxygen goes to the platinum of the negative electrode. (Sulphurous acid, carbonic oxide, carbonic acid). Carbonic acid is instantly decomposed into the gaseous lower state of oxidation, and into free oxygen, which gradually goes to the platinum. The carbonic oxide is gradually decomposed, by the oxygen leaving the carbon, and combining with the negative electrode.

Fig. 51 shows a very convenient arrangement for experimenting upon the spectra produced by different metals, comparing them with that produced by platinum. The metals, in the form of wires, are attached to screws, passing through clamps of vulcanite, which can be adjusted at any required height and angle by means of the spring tubes connecting them with the upright pillar. The wires on the left-hand clamp are permanently platinum, those on the

FIG. 51.

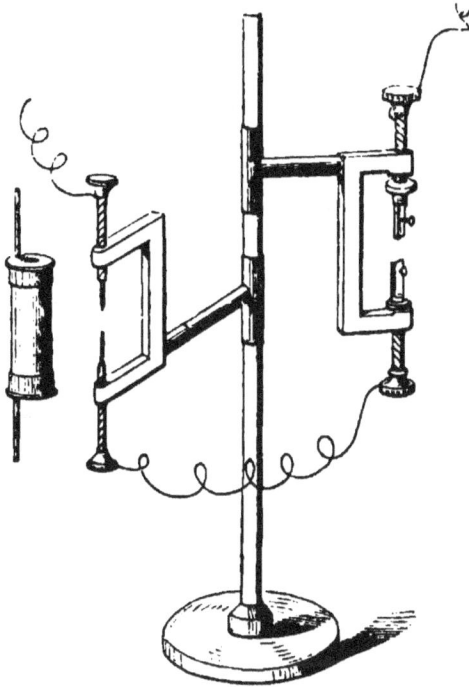

right-hand clamp may be of any other metal or metals; they are held by pincers, so that they may readily be removed and replaced by others. The two lower screws are metallically connected. The two upper are connected with the secondary terminals of the coil, and then with the Leyden jar, as in Ex. 8, p. 53. A brilliant discharge takes place simultaneously between the wires in each clamp, provided the distances be properly adjusted, and the apparatus being accurately arranged before the spectrum box, one spark is reflected through a prism, and the other is received directly through the slit; the two spectra immediately become apparent, one over the other, so that the peculiarities in each may be at once detected.

By employing the little capped glass tube, shown on the left-hand side of the figure, spectra may be obtained in various gases, the gas being passed through the tube while the discharge is taking place.

The Ozone Tube. It is well known that when electric sparks are taken between two conductors in atmospheric air a peculiar odour is developed. To the substance producing this odour the name *Ozone* has been given. It is supposed to be oxygen in an allotropic state, in which its chemical activity is greatly increased. It may be prepared 1stly by the action of clean moist phosphorus on atmospheric air; 2ndly, by the electrolysis of water acidulated with sulphuric acid; 3rdly, by passing electrical discharges through air or oxygen. A very ingenious little apparatus for the latter purpose is shown in Fig. 51.

FIG. 51.

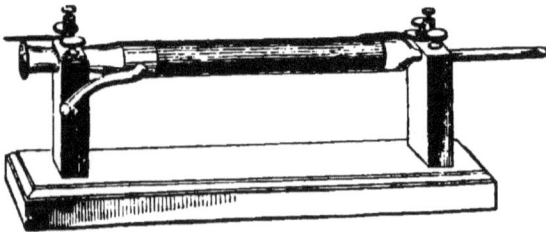

It consists of a glass tube about the size of an ordinary test tube, coated with tinfoil (or still better

silvered), and enclosed in an outer tube lined outside
with tinfoil. The two tubes are sealed together at
the neck of the outer one, and so adjusted that the
space between them shall be as narrow as possible.
At the projecting end of the inner tube is a brass
button, which is connected by a spring with one of
the binding screws on the frame of the apparatus,
which screw is to be connected with one of the
terminals of the secondary coil of an inductorium,
and the other with another binding screw in metallic
connection with the coating of exterior tube. The
apparatus is in fact a sort of slit Leyden phial, and
air or oxygen admitted through the lateral tube seen
in the figure becomes, during its passage through the
apparatus, powerfully ozonized. The air may be
driven through by means of a bladder or india-
rubber bag, or drawn through by an aspirator.

Ozone is a powerful oxidizing agent; it corrodes
organic matter; it bleaches indigo; it oxidizes the
metals converting even moist metallic silver into peroxide; but at the same time it seems, in some cases,
to act as a deoxygenant: thus it decomposes peroxide of hydrogen and peroxide of barium with the
evolution of inactive oxygen derived from both the
ozone and the peroxide. Schönbein regards ozone
as permanently *negative* oxygen, and he believes in
the existence of a permanently *positive* oxygen or
antozone; inactive oxygen he considers to be the pro-
duct of the union of ozone $\frac{-}{\text{o}}$ and antozone $\frac{+}{\text{o}}$.

New form of Thermo-Pile. The discovery of the
production of electricity by heating one of the junc-
tions of a metallic circuit, consisting of two metals
soldered together, was made by Professor Seebeck,
of Berlin, in 1821. The metals which give the
greatest amount of electro-motive force are bismuth
and tellurium, next comes bismuth and antimony,
and this latter metal, on account of its cheapness and

better conducting power, is generally substituted for Tellurium. The antimony is negative and the bismuth positive, the current going from the bismuth to the antimony across the junction. Numerous improvements on the original thermo-pile of Seebeck have been made by Nobili, Locke, Cumming, Dove, Van der Voort, etc. ; but the most efficient arrangement is that of Marcus, a representation of whose thermo-battery is shown in Fig. 52. It consists of thirty-six elements ; the negative bars, which are 6

FIG. 52.

inches long, being composed of—Antimony, 12 parts ; Zinc, 5 parts ; Bismuth, 1 part ; and the positive bars, which are 7 inches long, being composed of— Copper, 10 parts ; Zinc, 6 parts ; Nickel, 6 parts. The bars are ranged on a frame in the slanting position shown in the figure, the positive bar of the first pair being metallically connected with the negative bar of the second, and the two extreme bars connected with binding screws, which form the terminals of the battery. The upper ends of the bars are heated by a series of Bunsen's burners, the flames of which can be easily regulated.

Thermo-electricity is characterized by very feeble tension ; it can only therefore produce feeble chemical action. The battery above described will, however, though on so small a scale decompose water (feebly) give small sparks between iron points without the intervention of a coil; will enable the electro-magnet, shown in Fig. 3, to sustain 2 cwt.; and, when substituted for the voltaic battery with one of Ladd's 6-inch spark coils, will cause the production of sparks 1 inch long between the terminals of the secondary.

WORKS ON CHEMISTRY.

FOWNES' MANUAL OF CHEMISTRY. Edited by
H. BENCE JONES, M.D., F.R.S., and A. W. HOFMANN,
Ph. D., F.R.S. Ninth Edition, fcap. 8vo, cloth, 12s. 6d.

HANDBOOK OF VOLUMETRIC ANALYSIS; or, the
Quantitative Estimation of Chemical Substances by Measure.
By Francis SUTTON, F.C.S., Norwich. With Engravings.
Post 8vo, cloth, 7s. 6d.

THE USE OF THE BLOWPIPE. By Professors PLATT-
NER and MUSPRATT. Third Edition, 8vo, cloth, 10s. 6d.

THE FIRST STEP IN CHEMISTRY. By ROBERT
GALLOWAY. Third Edition, 310 pp., fcap. 8vo, cloth, 5s.

By the same Author,

THE SECOND STEP IN CHEMISTRY ; or, the Stu-
dent's Guide to the higher Branches of the Science. With
Engravings. Fcap. 8vo, cloth, 10s.

By the same Author,

MANUAL OF QUALITATIVE ANALYSIS. Fourth
Edition, post 8vo, cloth, 6s. 6d.

By the same Author,

CHEMICAL TABLES. On Five Large Sheets, for Schools
and Lecture-rooms. Second Edition, 4s. 6d. the Set.

PRACTICAL CHEMISTRY, INCLUDING ANALYSIS.
With numerous Illustrations on Wood. By JOHN E. BOW-
MAN. Edited by C. L. BLOXAM, Professor of Practical
Chemistry in King's College, London. Fifth Edition, fcap.
8vo, cloth, 6s. 6d.

By the same Author,

MEDICAL CHEMISTRY. With Illustrations on Wood.
Fourth Edition, fcap. 8vo, cloth, 6s. 6d.

NOTES FOR STUDENTS IN CHEMISTRY: being a
Syllabus of Chemistry and Practical Chemistry. By ALBERT
J. BERNAYS, Professor of Chemistry at St. Thomas's Hos-
pital. Fourth Edition, Revised, fcap. 8vo, cloth, 3s.

INSTRUCTION IN CHEMICAL ANALYSIS. By C.
REMIGIUS FRESENIUS. Edited by LLOYD BULLOCK.
QUALITATIVE. Sixth Edition, 8vo, cloth, 10s. 6d. QUAN-
TITATIVE. Fourth Edition, 8vo, cloth, 18s.

JOHN CHURCHILL AND SONS, NEW BURLINGTON-STREET.

HARDWICH'S PHOTOGRAPHIC CHEMISTRY.

Seventh Edition, thoroughly revised by GEORGE DAWSON, M.A., Lecturer on Photography, and E. A. HADOW, Demonstrator of Chemistry, in King's College, London. Fcap. 8vo, cloth, 7s. 6d.

" In selecting the two gentlemen whose names are appended as editors the publishers have shown much discrimination. It is satisfactory to us, and will no doubt be so to the public, to know that no profane hands have meddled with a work which all regard as belonging to the classical literature of photography."—*The British Journal of Photography.*

8vo, cloth, £2 10s.

AN EXPOSITORY LEXICON OF 50,000 SCIENTIFIC TERMS,

ANCIENT AND MODERN,

Including a Complete Medical and Medico-Legal Vocabulary, and presenting the Correct Pronunciation, Derivation, Definition, and Explanation of the Names, Analogues, Synonymes, and Phrases (in English, Latin, Greek, French, and German), employed in Science and connected with Medicine.

By R. G. MAYNE, M.D.

This Lexicon is suited to the requirements of every educated gentleman. It embraces the correct pronunciation, derivation, definition, and application of the names, analogues, synonymes and phrases (in English, Latin, Greek, French, and German), connected with Medicine, and employed in Anatomy, Animal Pathology, Astronomy, Botany, Chemistry, Comparative Anatomy, Conchology, Crystallography, Entomology, Geology, Geography, Geometry, Ichthyology, Materia Medica, Medical Jurisprudence, Medicine, Microscopy, Mineralogy, Natural History, Natural Philosophy, Nosology, Obstetrics, Ornithology, Pathological Anatomy, Pathology, Pharmacy, Phrenology, Physiology, Surgery, Trigonometry, and Zoology.

In one volume of 1,400 pages, with Engravings, Fourth Edition, greatly enlarged, 28s.

COOLEY'S CYCLOPÆDIA

OF

PRACTICAL RECEIPTS AND PROCESSES.

Being a General Book of Reference for the Manufacturer, Tradesman, Amateur, and Heads of Families.

From the "TIMES," *Nov.* 16, 1864.

" A much improved edition. It has become a standard work, not only as a supplement to the pharmacopœias, but also as a book of reference in connexion with the arts, manufactures, and trades."

JOHN CHURCHILL AND SONS, NEW BURLINGTON-STREET.

CATALOGUE

OF

Optical, Mathematical, and Philosophical

INSTRUMENTS.

MANUFACTURED AND SOLD BY

WILLIAM LADD,

11 & 12, BEAK STREET, REGENT STREET,

LONDON, W.

Microscope and Philosophical Instrument Manufacturer,

BY APPOINTMENT TO

THE ROYAL INSTITUTION OF GREAT BRITAIN;

THE GOVERNMENT SCHOOL OF MINES;

THE WAR DEPARTMENT;

THE EAST INDIA GOVERNMENT;

HER MAJESTY'S COMMISSIONERS OF NATIONAL EDUCATION;

THE GOVERNMENTS OF THE BRAZILS AND NETHERLANDS;

THE UNIVERSITIES OF OXFORD, CAMBRIDGE, LONDON, ETC.

————————◆————————

1866.

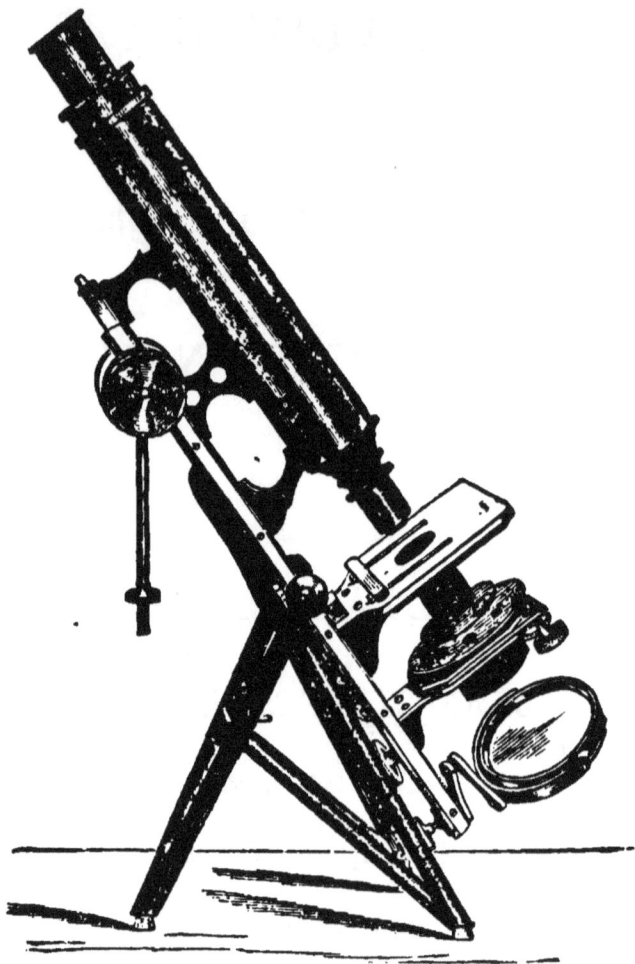

CATALOGUE.

MICROSCOPES.

	£	s.	d.
LARGE SIZE COMPOUND MICROSCOPE, of very superior workmanship and great solidity, the stage having 1-inch motion; plain and concave mirrors, fine adjustment (100 turns to the inch), secondary stage for holding achromatic condenser, spotted lens, &c., to which is applied the horizontal and vertical adjustments for insuring the perfect centricity of all its parts	18	18	0
Ditto, with binocular arrangement	26	0	0
Apparatus for the above:—			
Parabolic Condenser	1	10	0
Achromatic Condenser	5	10	0
Spotted Lens	0	15	0
Condenser on Brass stand	1	0	0
Polariscope, with selenite stage	2	5	0
Camera Lucida	1	0	0
Animalcule Cage	0	6	0
Extra deep eye-piece	0	15	0
Quarter-inch and 1 and 2-inch object-glasses of large angular aperture	7	7	0
One-eighth-inch object-glass	7	7	0
Mahogany Cabinet, with box for apparatus	2	15	0

LADD'S IMPROVED COMPOUND MICROSCOPES,
For which the Prize Medal was awarded, 1862.

	£	s.	d.
LADD'S IMPROVED COMPOUND MICROSCOPE:—			
With magnetic stage and two eye-pieces	9	0	0
Ditto, with mechanical stage, having rectangular movements	10	0	0
Ditto, with Binocular arrangement	16	0	0
Quarter-inch object-glass	4	4	0
1 and 2-inch object-glass, combined	3	3	0
Condenser on stand	0	18	0
Spot-glass for dark ground illumination	0	12	0
Polariscope	1	15	0
Animalcule Cage	0	6	0
Stage Forceps	0	5	0
Mahogany Case . . . . . . £1 10 0 to	2	10	0

*The above is strongly recommended in the "The Microscope,"
by Dr. Carpenter, page 81; and in "The Microscope; its History,
Construction, and Application," by Jabez Hogg, page 164.*

		£	s.	d.
COMPOUND ACHROMATIC MICROSCOPE, with moveable stage, having ¾ of an inch motion in rectangular directions, with sliding and revolving object-holder, two eye-pieces, double mirror, fine adjustment, diaphragms	8 0 0			
½-inch and 1-inch object-glasses	2 15 0			
Condenser on brass stand	0 18 0			
Animalcule Cage	0 6 0			
Mahogany Cabinet	1 10 0			
		14	9	0
Ditto, with Binocular arrangement		12	0	0
MICROSCOPE, designed by the late Geo. Jackson, Esq, in which the compound body, stage, and sub-stage are fitted in a dove-tailed slide running from top to bottom of the instrument, with improved magnetic stage, and eye-piece, in mahogany case		5	0	0
The above with ¼ and 1-inch achromatic object-glasses, and animalcule cage and forceps		8	0	0
EDUCATIONAL MICROSCOPE, with sliding stage, eye-piece, achromatic object-glasses, ranging from 30 to 300 diameters, condenser, animalcule cage, forceps, and mahogany case		5	5	0
EDUCATIONAL COMPOUND MICROSCOPE, with set of three achromatic object-glasses, eye-piece, and forceps, in mahogany case, with drawer		3	10	0
LADD'S AQUARIUM AND SEA-SIDE MICROSCOPE. The stage and mirror can readily be removed from the stand, so that the object-glass may be brought to bear upon the aquarium, and to follow an object with facility		4	0	0
PROFESSOR QUEKETT'S portable Dissecting Microscope, with drawer		2	10	0
Compound body for ditto, making it a portable Sea-side Microscope		1	0	0
Ladd's Dissecting Microscope, for botanical and other purposes		0	15	0
Condenser on stand for ditto, and can be used as a Microscope		0	6	0
Coddington Lens, of high magnifying power, very useful for opaque objects, mounted in ivory, German Silver, or silver 4s. to		5	0	0
Stanhope Lens, in various mountings . . . . from		0	2	6
Cloth Microscopes or Linen Provers, to fold for the pocket, 2s. to		0	4	6
Set of three lenses, for the pocket . . from 3s. 0d. to		0	7	6

MICROSCOPIC APPARATUS.

	£	s.	d.
Camera Lucida, for taking drawings of objects . . 15s. and	1	0	0
Neutral-tint Glass, for the same purpose	0	7	6
Erecting Glass, for dissecting with the Compound Microscope .	0	15	0
Side Reflectors, for illuminating opaque objects	0	18	0
Apparatus for Polarisation of Light . . . £1 5s. to	2	5	0
Achromatic Condensor, for transparent objects . £2 10s. and	5	10	0
Condensing Lens, on brass stand . . . . 18s. and	1	0	0
Parabolic Condenser, for dark ground illumination £1 5s. and	1	10	0
Spotted Lens, for low powers, by which a perfect black field is obtained . . . . 7s. 6d. to	0	15	0
Micro-Spectroscope, arranged for direct vision, and to shew two spectra at the same time, in the form of an eye-piece .	5	0	0
Compressorium	0	15	0
Extra Eye-pieces . . . . 12s. and	0	15	0
Animalcule Cage . . . . 3s. to	0	6	0
Glass Micrometers, for measuring the diameter of various objects, 100ths and 1000ths	0	5	0
Micrometer, mounted in brass frame, with screw adjustment fitted to eye-piece of Microscope	1	0	0
Slips of Glass, 3 inches by 1 . . per packet of 3 dozen	0	2	6
Glass Circles for Covers . . . . . per ounce	0	5	0
Ditto Squares „ . . . . . . „	0	4	0
Canada Balsam, pure Glycerine, Dean's Gelatine, Asphalt, Gold Size, &c. . . . . per bottle, 1s. to	0	1	6
Machine for cutting Sections of Wood . . 7s. 6d. to	1	1	0
Turntable for building up cells and varnishing the edge of covers	0	7	6
Brook's Double Connecter	1	10	0
Valentin's Knife, in case	0	16	0
Set of Microscopical Dissecting Instruments, in case . . .	1	10	0
Maltwood's Finder	0	7	6
A large assortment of Microscopic objects, sections of teeth, bone, &c.	0	1	6
Insects, Infusoria, and Vegetable Structures . . 1s. to	0	1	6
Anatomical Injected Preparations . . . . from	0	2	6
Microscopic Photographs	0	1	6
Mahogany Cabinets for holding 264 objects, and place for apparatus	1	7	6
Mahogany Cabinets, with glass doors, for Microscope objects from £2 2s. to	5	5	0
Glass Trough, for fixing fish, &c. . . . . 2s. 6d. to	0	5	0
Frog Plate	0	10	0
Glass Dissecting Trough	0	4	0
Dr. Beale's Cabinet, for Chemical Analysis, containing the following:—Platinum foil, test tubes, pipette, urinometer, graduated glass measure, spirit lamp with wire ring, watch glasses, glass slides, thin glass covers, and 8 re-agents in glass bottles, with capillary orifices	1	5	0
Injecting Syringe, with three jets and stopcock . . 12s. and	0	15	0
Improved Gas Lamp, with bath and plate for mounting objects	1	15	0
Diamond for writing on glass . . . . . 5s. to	0	7	0
Ditto, for plate or window glass	0	18	0
Apparatus for cutting thin glass circles	1	7	6

ACHROMATIC OBJECT-GLASSES

FOR MICROSCOPES.

Object Glasses.	Angular Aperture.	Price.	Magnifying Power with the various Eye-Glasses.		
			A	B	C
		£ s. d.			
2-inch	15 degrees	2 10 0	20	30	40
1½ „	20 „	2 15 0	40	15	70
1 „	15 „	1 11 6	60	80	100
1 & 2 „	25 „	3 3 0	„	„	„
½ „	65 „	4 4 0	100	130	180
¼ „	95 „	4 4 0	220	350	500
⅙ „	135 „	6 0 0	320	510	700
⅛ „	150 „	7 7 0	400	570	900

		£ s. d.
W. LADD SUPPLIES THE FOLLOWING WORKS.—		
Quekett's Practical Treatise on the use of the Microscope . .		1 1 0
The Microscope, by Dr. Carpenter		0 12 6
How to work with the Microscope, by Dr. Lionel Beale . .		0 5 6
Half-hours with the Microscope, by Dr. Lankester . . .		0 2 6
Hogg on the Microscope		0 6 0

TELESCOPES.

EQUATORIAL, with 4-in. object-glass, 5 ft. focal length; 6 astronomical eye pieces; diagonal and transit eye-pieces; double parallel-wire Micrometer, with 4 eye-pieces; illuminating apparatus; declinating and hour circles, graduated on silver, with Vernas and Microscope. Supported on stout iron pillar 110 0 0

Clock-work for the above 20 0 0

FOUR-FT. ASTRONOMICAL TELESCOPE, 3½-inch object-glass, 5 eye-pieces, and brass tripod, with horizontal and vertical movements, packed in mahogany case . . . 31 10 0

3-ft. 6-in. ditto, 3 eye-pieces, 2¾-in. object-glass, in mahogany case 21 0 0

3-ft. ditto, 3 eye-pieces, 2¾-in. object-glass, in mahogany case 15 0 0

2-ft. 6-in. ditto, 2 eye-pieces 10 0 0

Tripod Garden Stand, suitable for either of the above   35s. to 2 10 0

21-in. Navy Telescope 2 5 0

18-in. ditto 1 16 0

15-in. ditto 1 10 0

Day and Night Telescopes . . . . . . from 1 10 0

Pocket Telescopes of every description and of best quality.

		£ s. d.
Achromatic Stereoscope . . . . . . from		1 10 0
Revolving Stereoscope, to hold two dozen glass slides . .		3 10 0
A large assortment of Glass Stereoscopic Slides from 4s. 6d. to		0 6 6
Wheatstone's Reflecting Stereoscope . . . . from		2 0 0
Ditto Pseudoscope		1 10 0

OPERA AND RACE GLASSES,

With achromatic eye-pieces, with ivory, pearl, tortoiseshell, enamelled, or leather mounts.

MAGNIFYING GLASSES,

For viewing prints and paintings.

SPECTACLES AND EYE-GLASSES

In every variety of mounting.

Theodolites, Levels, Compasses, Sextants, and Quadrants.

MAGIC LANTERNS, DISSOLVING-VIEW APPARATUS, ETC.

	£	s.	d.
Magic Lanterns . . . . from 7s. to	1	10	0
Phantasmagoria Lantern, with best Argand Lamp, 3-in. condensers, in a deal case	3	0	0
Ditto, with 3½-inch condensers	4	0	0
Ditto, with Microscope, 3½-inch condensers, microscopic objects and water trough, packed in case	5	10	0
Oxycalcium Lantern, for exhibiting 3-inch pictures on a screen 10 feet diameter, with apparatus complete	9	0	0
Dissolving-View Apparatus, with 3½-inch condensers, for showing 3-inch pictures	9	0	0
Ditto, with Oxycalcium apparatus complete	15	0	0
Microscope for ditto . . . . from	1	0	0
Dissolving-view Apparatus, with oxy-hydrogen lime light, 4-inch condensers, 2 gas bags, pressure boards, tubing. gas generators, purifier, and copper retort complete .	30	0·	0
Dissolving-view Apparatus combined in one lantern, with stopcock arrangement for producing dissolving effect, with gas bags, &c., complete	24	0	0
Improved Oxy-hydrogen Microscope, with 3 powers and mahogany lantern	7	10	0
Ditto, with Oxy-hydrogen jet, gas bags, pressure boards, tripod stand, and apparatus for making gas, &c.	30	0	0
Apparatus to show magnetic curves	0	12	0
Clockwork movement for revolving the lime cylinders . each	2	5	0
Opaque screens. 8-feet square, on rollers			
Ditto 10-feet „			
Ditto 12-feet „			
One dozen Lime Cylinders, in sealed bottle	0	3	0
Dissolving-views, separate or in sets, to illustrate Astronomy, Geology, Natural History, &c.			
The whole and parts of insects, sections of woods, &c., prepared specially for the Oxy-hydrogen Microscope, each, 2s., 2s. 6d. and	0	3	0

APPARATUS FOR ELECTRIC LIGHT, POLAR-ISATION, &c.

	£	s.	d.
Ladd's improved Electric Lamp	10	10	0
Lantern for ditto, with Condensers	6	10	0
Holmes' Improved Electric Lamp for Lighthouses, &c. . .	18	0	0
Parabolic Reflectors, thickly silvered	2	2	0
Revolving Diaphragm for Lantern	0	15	0
Rectangular ditto	1	10	0
Mahogany Box, containing Plyers, extra pair of Carbon Holders, Forceps, and 6 ft. of Carbon Points	1	0	0
Best Carbon Points . . . . per feet	0	1	6
Microscope for the Electric Light, with best Achromatic Objectives	10	0	0
Improved Polariscope for above	12	10	0
Prisms of heavy Glass, on Stand . . . from £2 to	3	10	0
Ditto of Quartz . . . . from £3 to	5	0	0
Polyprisms of Six Glasses of various density, on Stand . .	3	10	0
Bottle Prisms of Bisulphide of Carbon . . from 12s. to	0	18	0
Prism with movable sides, on Brass Stand, with adjusting Screws	6	0	0
Adjustable Stand for Prisms	0	7	6
4-inch Condenser, on Stand, for focusing image on screen . .	1	10	0
Bunsen's burners . . . . . from 4s. 6d. to	0	10	0
4 Sets of 10 Grove's Batteries for Electric Light . . .	20	0	0
Biot's Reflecting Polariscope, with Apparatus	5	5	0
Tourmaline Polariscope, to illustrate the system of coloured rings, in crystals, &c.	2	2	0
Ditto, Pincers for ditto . . . . . from	1	10	0
Selenites of various devices . . . . . from	0	5	0
Specimens of unannealed Glass, of various shapes . from	0	3	0
Apparatus to shew Newton's rings	1	1	0
2 Prisms to illustrate Achromatism on stand	1	10	0
2 Glasses ruled with fine lines to shew Iridescence . from	0	10	0
Concave and Convex Mirrors			
Spheres of Iceland Spar	2	0	0
Plates of Arragonite, Quartz, Topaz, Nitre, Calc-spar, Borax, &c., for exhibiting the coloured rings . . . from	0	5	0
Apparatus to show the polarising structure, communicated to glass by pressure . . . . from 7s. 6d. to	1	1	0
Rectangular Prisms of Glass for ditto			

Rhombs of Iceland Spar.	Polariscopes fitted to Microscopes.
Polished Plates of Tourmaline.	Selenite Plates.
Double and Single Image Prisms.	Gutta Percha covered Wire, per
Polarising bundles of Crown Glass.	yard, 3d. to 6d.

SPECTROSCOPES AND APPARATUS.

	£	s.	d.
Apparatus for Spectrum Analyses with four Prisms, very superior Spectroscope with two Prisms, graduated circle, brass support &c., packed in mahogany case	15	0	0
Spectroscopes with one Prism, Photographed Micrometer Scale, &c.	6	0	0

	£	s.	d.
Ladd's Universal Spectroscope with two Prisms packed in case.	5	5	0
The above can be used in any position either with or without Microscope.			
Pocket Spectroscope for direct vision	3	3	0
Micro-Spectroscope arranged for direct vision, and to shew two Spectra at the same time in the form of an eye-piece .	5	0	0
Bunsen's Burners, for the above . . . . . . from	0	4	6
Apparatus consisting of a Brass Stand, with two Insulated Dischargers for obtaining the Spectra of Metals, by means of the Induced Spark, with tube for Gases (Fig. 51) . .	1	15	0

LADD'S

IMPROVED INDUCTION COILS & APPARATUS.

For which Prize Medal was awarded at the Exhibition, 1862.

INDUCTION COIL, to give 1¼-inch spark in air . . .	10	10	0
Ditto ditto 2¼-inch ditto . . .	12	12	0
Ditto ditto 4-inch ditto . . .	15	15	0
Set of 5 Grove's Batteries, with platina plates 5×2¼-inch, in tray	2	15	0
Ditto ditto 6½×3-inch . .	4	10	0
Apparatus consisting of glass tube with two platinum terminals, with brass plate and glass receiver to fit upon air-pump, for experiments with Torricellian vacuum (Fig. 23.) . .	1	10	0
Gassiot's Torricellian Vacuum Tube, for broad, cloudy stratification (Fig. 26.)	1	10	0
Ditto, packed in case	1	16	0
Uranium Glass Tube, mounted on Stand, with stopcock £1 1s. to	2	2	0
Glass Tubes for showing the Aurora Borealis, fitted with stopcock, and capable of being charged with various gases, from 2 to 6 ft. long, and 1 to 4 inch diameter (Fig. 21) . .			
Apparatus for showing the rotation of a spark round an electro-magnet (Fig. 19.)	3	10	0
Ditto (Fig. 20.)	1	10	0
Bar Electro-magnet, for experimenting with the electric spark . . . . from	0	18	0
Egg-shaped Glass, with stopcock and sliding wire (Fig. 22) .	2	5	0
GASSIOT'S CASCADE, permanently exhausted (Fig. 16) .	3	10	0
Gassiot's Revolving Star (Fig. 48) . . . £3 3s. to	4	10	0
Geissler's Sealed Vacuum Tubes. — These tubes have been charged with the various gases, and then exhausted to the utmost, and hermetically sealed.			
Carbonic Acid Vacuum Tube, with stick of caustic potash at one end (Fig. 29) . . . . . £1 5s. and	1	10	0
Vacuum tubes composed of two or more distinct vacuums, showing a variety of colours according to the gases contained			
Double Garland Tube (Fig. 34)	2	0	0

	£	s.	d.
Single Garland Tube (Fig. 35)	1	10	0
Ditto ditto with letters (Fig. 36) . . each	2	2	0
Vacuum Tubes (Figs. 37, 38, 40, & 42) . . . . each	1	5	0
Ditto (Figs. 39, 43, & 44) . . . . . each	1	10	0
Ditto (Fig. 41) . . . . . . each	1	1	0
Ditto (Fig. 47) . . . . . . 15s. to	1	1	0
Ditto of Uranium Glass . . . . . 15s. to	1	10	0

These tubes vary in size, and consequently in price.

	£	s.	d.
Vacuum Tubes, charged with various Gases for Spectrum Analysis (Fig. 45)	0	10	0
Gassiot's support for six ditto; the tubes become successively illuminated when in front of the Spectroscope . . .	2	0	0
Vacuum Tube, for surgical purposes (Fig. 46)	1	1	0

Vacuum Coronets.
Ditto, Miners' Lamps.
Ditto, to show extraordinary effect when placed between the poles of a powerful electro-magnet. (Figs. 24 and 25).

A variety of other Tubes always in stock.

	£	s.	d.
Udiometer to be used with Inductorium . . . from	0	5	0
Apparatus for producing Ozone in large quantities by aid of the Induction Coil. (Fig. 52)	1	1	0
Uranium Glass Vessel for showing fluorescence	0	5	0
Block of Uranium Glass, in mahogany case . . from	1	5	0
Glass Tubes, containing Becquerel's Phosphorescent Powders in sets of six . . . . 15s. and	1	5	0
Revolving Colour Disc, for showing white light, also for proving that the Induction Spark is not continuous . . .	0	7	0
Spotted Jars . . . . 7s. 6d. to	2	2	0

W. Ladd has been appointed Sole Agent for Geissler's Vacuum Tubes and Chemical Apparatus.
The above figures refer to " Treatise on the Induction Coil," by Dr. H. M. Noad, F.R.S., &c., with 40 illustrations, price 3s.—W. Ladd, Beak St.

APPARATUS FOR FRICTIONAL ELECTRICITY.

	£	s.	d.
36-inch, Plate Electrical Machine, mounted upon the best principle with Electrometer attached	22	0	0
30-inch, ditto ditto	15	15	0
24-inch, ditto ditto	10	10	0
18-inch, ditto ditto . . . £7 and	8	5	0
15-inch, Plate Electrical Machine	5	0	0
12-inch, ditto ditto without Electrometer . .	4	0	0
9-inch, ditto ditto ditto . . .	3	0	0

Cylinder Electrical Machines made to order.

	£	s.	d.
Improved form of Electrometer, with Condenser (very delicate)	3	3	0
Bohnenberger's Electroscope			

	£	s.	d.
Quadrant Electrometer	0	7	0
Cuthbertson's Self-acting Electrometer	1	18	0
Bennet's Gold Leaf Electrometer . . . . 14s. to	1	10	0
Medical Jar	0	5	0
Henley's Universal Discharger, with Press and Table £1 5s. and	2	0	0
Electrical Cannon, with brass carriage	0	18	0
Electrical Flask, with cap and valve .	0	6	0
Electrical Pistol	0	5	0
Luminous Conductor .	0	18	0
Electrical Sportsman . . . . 15s. and	1	1	0
Egg-stand .	0	7	6
Egg-shape Glass, with stopcock and sliding wire for showing light in vacuo	1	15	0
Hand Spiral . . . . . . 3s. 6d. and	0	5	0
Set of five Spirals	1	8	0
Luminous Words, in frame .	0	10	6
Revolving Spiral, on stand .	0	8	6
Image Plates with brass stand . . . . 8s. 6d. and	0	10	6
Dancing figures made of pith	0	1	0
Electrical See-saw	0	12	0
Pith-ball Stand .	0	5	6
Pith Balls . . . . . . per dozen, 9d. and	0	1	0
Carved Head with hair . . . . .. 3s. 6d. and	0	5	6
Diamond Jar . . . . . . 7s. 6d. and	0	12	0
Bucket and Syphon .	0	5	0
Electrical Orrery	0	7	6
Sturgeon's Apparatus for firing Gunpowder, &c.	0	9	0
Insulated Brass or Wood Stands			
Electrical Spider	0	1	0
Electrical Obelisk	0	7	0
Thunder House .	0	7	0
Fire House .	0	12	6
Gamut of Bells .	1	10	0
Set of three Bells	0	7	0
Insulated Stool .	0	12	0
Jointed Discharger . . . . . 7s. 6d., 9s., and	0	10	6
Discharging Rods	0	3	6
Apparatus for showing the falling star	0	15	0
Electrical Cylinders			
Glass Handles and Legs			
Brass Chain . . . . per yard	0	0	4
Amalgam . . . . per box	0	1	0
Circular Glass Plates for Electrical Machines			
Ditto Ebonite discs for ditto			
Conductors			
Leyden Jars . . . . from	0	3	6
Brass Balls . . . . from	0	0	9
Electrophorous . . . . . . from	0	10	0
Glass Jar, with movable tin coatings .	0	10	6
Set of Electrical Apparatus for Educational purposes, consisting of the following articles:—Plate Electrical Machine, Leyden Jar, Pith-ball Stand, Jointed Discharger, Hand Spiral, Brass Clamp, Head of Hair, Amalgam, and Chain, packed in coloured deal case, with lock and key	5	5	0

VOLTAIC AND MAGNETIC APPARATUS.

	£	s.	d.
WILDE'S Magneto-Electric Machine (Fig. 8). This Machine can be used for a variety of experiments, and is an excellent substitute for a Voltaic Battery	21	0	0
Medical Galvanic Coil, of improved construction, which can be regulated so as to apply it either to an infant, or to the most obstinate cases, in mahogany box	4	4	0
Improved Medical Coil, with sulphate of mercury battery, complete in the form of a book (Fig. 9).	4	4	0
Medical Galvanic Machine, in mahogany box	3	3	0
Ditto ditto small size	2	2	0
Improved Magneto-electric Machines, for medical purposes, in mahogany case (Fig. 5)	2	5	0
Electro-magnet . . . . from	0	6	0
Mahogany support for ditto	0	5	0
Powerful Electro-magnet on stand	2	0	0
Ditto, very powerful with double wires, movable coils, mahogany stand, &c., (can be used for Dia-magnetic experiments) .	10	0	0
Rheostat . . . . from	5	10	0
Rheochord	1	15	0
Wheatstone's Bridge			
Galvanometer . . . . from	0	10	0
Galvanometers with Astatic Needle, with levelling screws and glass shade . . . . from	1	8	0
Ditto (very delicate) (Fig. 4) . . . . from £3 3s. to	4	4	0
Barlow's Wheel	0	10	0
Sturgeon's Disc, to go with the above	0	3	0
Oersted's Experiment . . . . . 7s. 6d. and	0	10	6
Dipping Needle, on brass stand with divided arc . .	1	1	0
Ladd's Improved Electro-motive Engine, will raise 30 lbs., size of stand, 7 by 5½ inches	7	0	0
Apparatus for showing the rotation of Electro-magnet between the poles of a soft horse-shoe	0	10	0
Richie's Electro-magnetic Apparatus, consisting of horse-shoe-magnet, on stand, with levelling screws, Armature, Ampere's Bucket, wire frame, helical coil, and two flood cups . .	2	10	0
Marsh's Vibrating Suspended Wire	0	7	0
Working Model of Telegraph, by which sentences may be transmitted	3	10	0
Morse's Printing Telegraph, consisting of receiver, relay, and key	10	10	0
Apparatus for Ringing a Bell by Electro-magnetism . . .	1	1	0
Apparatus for Ringing a Bell, of improved construction . .	2	2	0
Double Commutater	0	15	0
Commutater (the same as used with the Induction Coil) .	1	1	0
Decomposition Apparatus . . . . . 7s. 6d. to	3	3	0
V Tube for Decomposition	0	7	6
Faraday's Voltameter	5	5	0
Melloni's Thermo-electric Battery of twenty-five pairs of Antimony and Bismuth Bars	2	0	0
Ditto (very delicate), with conical reflector	5	5	0
Large Thermo-Electric Battery (Fig. 53)	8	8	0

	£	s.	d.
Ampere's Stand (improved form), by which the experiments relating to the mutual attraction and repulsion of electrical currents can be shown	2	0	0
Smee's Battery (pints)	0	7	6
Ditto (quarts)	0	10	0
Set of six Smee's Batteries, in vulcanised cells, in mahogany trough	4	0	0
Ditto, with adjustments for raising it out of the cells; may be arranged for either quantity or intensity	4	10	0
Grove's Platinum Battery	0	12	0
Set of eight of Grove's Batteries in tray	5	5	0
Set of ten ditto, in tray	6	0	0
4 sets of ditto, for electric Light	24	0	0
Improved Coke Batteries, in glass or stoneware cells			
Brass Clamps for Batteries			
Glass and Porous Cells			
Platinum Foil and Wire . . . . . . per oz.	1	12	0
Platinised Silver . . . . ,,	0	10	0
Amalgamating Zinc Plates . . . . . per lb.	0	1	0
Copper Wire of all sizes, covered with silk or cotton ,,			
Set of Electro-Magnetic Apparatus for Educational purposes, consisting of an Electro-magnetic Coil Machine, Smee's Battery, Galvanometer, Richie's Experiment, Oersted's ditto, Electromagnet and mahogany stand, Barlow's Spur-wheel and Permanent Magnet, packed in coloured deal case with lock	5	5	0

ABEL'S FUZES,

FOR FIRING MINES AND CANNON BY MAGNETIC ELECTRICITY.

W. L. is appointed sole manufacturer of these by order of the Secretary of State for War.

	£	s.	d.
Experimental Fuzes . . . . per doz.	0	2	0
Blasting ,, ... ,,	0	2	9
Cannon	0	3	9
Magneto-Electric Exploder, in mahogany case, with two keys	7	0	0
Ditto, very powerful	17	17	0
Ditto, with six keys, Fig. 6	18	18	0
Induction Coil, specially adapted for blasting purposes where a large number of fuzes are required to be fired simultaneously, in strong oak case	10	10	0
G. P. Insulating Wire, for connections . per 100 yards from	1	0	0
Ebonite Electrical Machine, or Austrian Exploder . . .	13	0	0
Oak case for ditto	2	2	0

DRAWING INSTRUMENTS, &c.

	£	s.	d.
Sets of Drawing Instruments, for youths 5s. 6d., 7s. 6d. and	0	10	6
Ditto, in mahogany cases	1	1	0
Ditto ditto	2	2	0
Ditto ditto, German silver	2	10	0
Ditto ditto, very superior ditto	3	10	0
Architect's case of ditto, for the pocket	2	2	0
Ditto ditto, best make, German silver . . .	3	10	0

A complete set of Mathematical Drawing Instruments, of the
best construction, with proportional compasses, rule, and
scales . . . . from £10 10s. to **21 0 0**

Pentagraph, for copying, reducing, and enlarging plans, draw-
ings, maps, &c., with case, 18-in., £4 10s.; 24 in, £5 10s.;
30-in., **6 10 0**

		£	s.	d.
Drawing Pens	3s., 4s., and	0	6	0
Proportional Compasses, 6-inch	£1 10s. and	2	10	0
Engineers' Pocket Compasses		1	0	0
Ditto ditto, best make, German silver . . .		2	0	0

Bow, Pen, and Pencils.
Spring and Hair Dividers.
Spring Dividers, Pen and Pencil, the set 10s. 6d. to £1 5s. 6d.
Rolling Parallel Rules, 6-in., 6s.; 9-in., 7s.; 12-in., 8s. 6d.; 15-in., 11s.
Rolling Parallel Rules, brass, 1s. 6d. per inch.
Protractors.
Sectors.
Mahogany and Ebony T-squares.
Drawing Pins.

METEOROLOGICAL INSTRUMENTS.

		£	s.	d.
Standard Barometers	from	6	6	0
Pedestral Barometers, in mahogany, walnut, or rosewood frames,	£2 2s. to	7	7	0
Wheel Barometer	£1 15s. to	6	6	0
Board of Trade Marine Barometer, in case		5	0	0
Fitzroy's Sea-coast Barometer		4	0	0
Marine Barometer	£3 3s. to	5	5	0
Marine Barometer and Simpiesometer, in one instrument	from	5	0	0
Pocket Barometer		2	15	0
Barometer, with compensating tube and extended scale (very delicate)	£1 15s. and	2	10	0
Ditto ditto, Standard				
Aneroid Barometer	from	2	10	0
Ditto, for the pocket		3	0	0
Patent Mercurial Maximum Thermometers		0	15	0
Standard Thermometers	10s. 6d. to	1	1	0
Ditto, with spiral bulb, very delicate		1	10	0
Thermometers for registering extreme heat and cold .	from	0	8	6

	£	s.	d.
Ivory Thermometers, in leather case . . . . 4s. 6d. to	0	10	0
Chemical Thermometers, divided on glass . . . from	0	4	6
Botanical Thermometers, in tin case . . . . ,,	0	9	0
Air Thermometers . . . . ,,	0	10	6
Leslie's Deferential Ditto . . . . . . ,,	0	15	0
Boxwood Thermometers . . . . ,,	0	1	0
Wet and Dry Bulb Thermometer	2	2	0
Mason's Hygrometer	1	6	0
Robinson's Anemometer, for ascertaining the velocity of the wind	3	3	0
Lind's ditto	2	2	0
Electrometer, for Atmospheric Electricity	2	2	0
Rain-gauge, in japanned tin or copper . . . from	1	0	0

W. L. is sole Agent for Geissler's Chemical Thermometers.

PNEUMATIC APPARATUS.

	£	s.	d.
LADD'S SUPERIOR large size double-barrel Air-pump, with additional barrel for very accurate exhaustion, barometer, gauge, &c., on strong mahogany stool, 12-inch plate . .	35	0	0
Ditto, ditto, smaller size, for table, Fig. 17 . . £15 and	20	0	0
Auxiliary Mercury Pump for obtaining a perfect Vacuum (Fig. 18)	7	7	0
Grove's Pump, with 7-in. plate, mercurial gauge, and two clamps	5	0	0
Tate's Pump, with gauge, two clamps, and key . . .	4	4	0
Large size double-barrel Air-pump, with raised plate, 10 inches in diameter, gauge-plate, mercurial gauge, clamp and key	11	10	0
Second size double-barrel table Air-pump, with raised plate, 9 inches diameter, gauge-plate, gauge and key . .	9	10	0
Ditto, ditto, with plate 8 in. in diameter, on stand (not raised), with gauge-plate, gauge and key	8	0	0
Ditto, ditto, without gauge-plate, gauge and key . .	6	10	0
Third size double-barrel Air-pump, diameter of plate 6¼ in. .	4	4	0
Smaller size double-barrel Air-pump, diameter of plate 5¼ in. .	3	10	0
Small size single-barrel Pump, 3½-inch plate	1	0	0
No. 2, ditto, ditto, 4½-in. plate	1	10	0
,, 3 ,, ,, 5½ ,, ...	1	15	0
,, 4, ,, ,, sloping barrel, 6½-in. plate	2	2	0
Flat Brass Plate, with sliding wire . . . 10s. 6d. and	0	13	0
Exhausting or Condensing Syringes	0	7	0
Ditto ditto, in one instrument	0	10	0
Apparatus, consisting of Glass Cylinder and Piston, to show the effect of pressure upon gases	1	0	0
Fire Syringe	0	3	6
Bell Experiments . . . . . . 7s. 6d. and	1	0	0
Bacchus Experiment	1	8	0
Balloons of Goldbeater's skin, that will ascend with ordinary gas, 9-in. 1s.; 10½-in. 2s; 12-in. 2s. 9d.; 14-in. 3s. 6d.; pear shape . . . . each	0	5	0

	£	s.	d.
Hand and Bladder Glass	0	2	6
Lungs Glass	0	6	0
Large size Hemispheres	1	8	0
Middle size ditto	0	16	0
Small size ditto	0	12	0
Filtering Cup, with Brass Plate	0	6	6
Three-fall Guinea-and-Feather Apparatus	1	1	0
Two-fall ditto	0	15	0
Tall Glass Receiver for ditto . . . . 7s. 6d. and	0	10	6
Windmill (improved)	1	15	0
Double-Transferer	1	16	0
Single ditto . . . . 12s. and	0	18	0
Bladder Frame and Lead Weights	0	8	0
Copper Bottle, Beam, and Stand	2	5	0
Fruit and Taper Stand	0	3	0
Syringe and Lead Weights	0	8	0
Balance Beam and Cork Ball, with counterpoise weight . .	0	10	0
Torricellian Experiment	0	15	6
Ditto, having the Barometer fixed in the cap of glass receiver .	1	7	6
Glass Globe, with brass cap and stopcock for weighing air .	0	8	0
Leslie's Apparatus for freezing water	0	10	0
Breaking Squares	0	1	3
Wire Cage for ditto	0	4	6
Brass Stopcocks . . . . 2s. 6d. and	0	3	6
Apparatus for showing fountain in vacuo	0	12	0
Tall receiver for ditto	0	7	6

LADD'S EDUCATIONAL SET OF PNEUMATIC APPARATUS.

As supplied by him to the various Educational Societies, consisting of the following Articles :—

Single-barrel Air-pump and Receiver, Brass Clamp, Filtering cup for Mercury, Magdeburgh Hemispheres with handles, Bladder Frame and Weights, Guinea-and-Feather Apparatus, Fruit and Taper Stand, Hand and Bladder Glass, Single Transferer and Fountain Apparatus, Brass Pipe for ditto, Bell Experiment, Brass Syringe for instantaneous Light, Glass for Fountain and Guinea-and-Feather Apparatus, and Plate for Top of Fountain Glass, packed in coloured case, with lock and key . . . . .   6 .6 0

ACOUSTICS.

	£	s.	d.
Stand, with Organ bellows and sound-board, with holes for organ pipes, &c . . . . from	5	0	0
Set of eight organ pipes	2	2	0
Set of organ pipes to illustrate various methods of producing musical notes.			
Organ pipe with membranes and gas jets to illustrate the nodal points in a column of air	1	15	0

	£	s.	d.
Syrenes . . . . from £2 to	4	0	0
Perforated disc, mounted on revolving stand, with jet and mouthpiece, to illustrate the production of musical sound by regular and irregular impulses	2	10	0
Tuning-fork, mounted on sound case	1	10	0
Set of 4 smaller ditto, making a perfect chord	4	10	0
Ditto ditto, mounted on stand, to record vibrations upon smoked glass	4	10	0
Set of eight ditto, on sounding cases, forming the gamut . .	9	0	0
Set of 4 Forks, with reflectors for showing the curves produced by the composition of rectangular vibration . . .	10	10	0
Two stands for ditto	1	10	0
N.B.—This set shows the following figures. unison, second, third, fourth, fifth, sixth, seventh, and eighth, and intermediate semitones.			
Set of eight ditto, with reflectors, supports, and lamp ; also with support and sliding frame, for recording vibrations on smoked glass	28	0	0
Apparatus consisting of 2 vibration-springs fitted to lantern, by means of which, the whole of the above figures can be pro · jected on the screen	3	10	0
Large Tuning Fork with Electro-magnet to keep up constant vibration of strings.			
Whirling apparatus for showing notal points in vibrating strings.			
Whirling Table . . . . £2 and	4	10	0
Apparatus consisting of a bell, with sliding tube to augment sound	2	12	6
Wheatstone's Kaleidophone . . . . from 10s. to	1	10	0
Apparatus to illustrate normal or transverse undulations in a row of particles	4	14	6
The Monochord . . . . irom £2 2s. to	5	5	0
A long tube, with piston, for experiments on the reciprocation of sound and on multiple resonance	2	2	0
Wheatstone's Apparatus for proving the simple mode of vibration of a tube open at both ends	1	11	6
Long Vertical Gas Jet, with brass foot and stopcock, to produce musical notes in glass tubes	0	15	0
Apparatus for rotating Gas Jet in a Glass Tube, with multiplying wheel and stand	2	2	0
Willis's Tube for the production of vowel sounds £2 2s. and	3	3	0
Set of Membranous Apparatus to illustrate the production of the human voice	0	7	6
Stands for clamping-rods at the ends, or at one or more nodal points. to show their transverse vibrations.			
Large double brass clamp for holding plates	0	18	0
A series of Six Glass plates of different forms to illustrate the vibrations of elastic surfaces	0	9	0
Circular Metal Plate, on stand, to show ditto	0	18	0
Set of 3 ditto	2	5	0
Square Metal Plate, ditto ditto	0	18	0
Extra Large ditto ditto	1	1	0
Pounce Boxes . . . . each	0	1	0
A Square, Circular, and Triangular Frame, over which is stretched a delicate tissue, to show the vibrations of elastic membranes	0	15	0

	£	s.	d.
Apparatus for exhibiting the nodes of a bell consisting of a glass vessel over the rim of which is suspended a row of cork balls	1	18	0
Hopkin's Apparatus to show the interference of sound . .	0	10	6
Various Apparatus for showing that sonorous vibrations are always transmitted in the direction they were originally propagated.			
Various Apparatus to show the interference of sonorous undulations and the analogies between these and the interference of light.			
Trevyllian's Rocking Bar and Lead Weight	0	12	0
Long glass tube, with brass plate, to produce sound by the flow of water through a small aperture.			
Set of Steel Spirals, mounted on a sounding box, with hammer	2	2	0
Strong Violoncello Bows for vibrating various apparatus from	0	12	0
2 Circular Brass Plates separated by long brass rod . . .	0	15	0
Glass Rod, with square or circular disc attached to show the influence of the surrounding medium on the acoustic figures produced.			

MECHANICS, HYDRAULICS, &c.

	£	s.	d.
Educational Set of Mechanical Powers . . from £3 3s. to	20	0	0
Sets of Levers, various . . , . . from £1 1s. to	4	4	0
Sets of Pulleys, in frame, to show various arrangements . .	2	0	0
Single and Double Incline Planes, with rollers, from 10s. 6d. to	3	10	0
Various apparatus to illustrate the resolution and composition of force, the Equilibrium and Centre of Gravity of bodies, &c.			
Gyroscopes . . . . from 25s. to	5	5	0
Whirling Tables for demonstrating the laws of central forces.			
Atwood's Machines . . . . from	4	4	0
Apparatus to illustrate the laws of collision	2	0	0
Dissected Cones . . . . from	0	9	0
Geometrical Solids . . . . . . per set	0	10	0
Large set of ditto (2-inch cube)	1	10	0
Working Model of Bramah's Hydrostatic Presses from £5 to	25	0	0
Apparatus to illustrate the principle that fluids will rise to the same height . . . . from	0	5	0.
Tantalus Cup	0	10	0
Glass Syphon . . . . 2s. and	0	3	6
Glass Balloons, Divers, &c. . . . . . each	0	2	0
Ditto, ditto, with Tall Jar . . . . . . from	0	5	0
Model of Centrifugal Pump	3	3	0
Model of Lifting Pump, with glass barrel	0	18	0
Ditto of Forcing Pump	1	8	0
Model of Archimedes' Screw, with glass worm	2	2	0
Ditto of Undershot Wheel	1	15	0
Ditto of Overshot Wheel	1	15	0
Ditto of Diving Bell, with Force-pump	1	1	0

	£	s.	d.
Fountain Apparatus, consisting of strong metal vessels, stopcock, condensing syringe, and set of jets . . from	2	2	0
Philosophical Water Hammer	0	5	0
Woolaston's Cryophorus . . . . 4s. and	0	6	0

HEAT.

	£	s.	d.
Ferguson's Pyrometer	5	0	0
Daniell's ditto	5	0	0
Metal Ball and Ring	0	10	6
Set of 5 Balls of different metals, to illustrate their specific heat	0	10	0
Compound Bar of Iron and Brass	0	10	0
Metal Bar and Gauge	0	6	0
Iron and Brass Bars, supported on mahogany stand, with connections for battery, to illustrate the different expansion of metals.			
Apparatus for showing the force exerted by the contraction of solids . . . . , .	0	15	0
Iron Bottles to show expansion of water at freezing point, and Bismuth on cooling . . . . . . each	0	2	6
2 Metal Bars on stand, with spirit lamp, &c., to show expansion	0	18	0
Ditto, with gas burner	1	5	0
Metal Bar, on stand, to show conduction, with cups, balls, and lamps	0	8	6
6 different Metal Bars, emanating from one centre, to show ditto . . . . 5s. 6d. and	0	7	6
Faraday's Convection Apparatus	0	15	0
Glass Globe and Bucket, to illustrate the circulation of heated water.			
Fire Balloons.			
Parabolic Reflections for radiation and reflection . per pair from £2 2s. to	12	12	0
Iron Ball and Stand for ditto	0	5	0
Leslie's Thermometer	0	15	0
Pewter or Tin Cubes for radiation . . . from 2s. 6d. to	0	12	0
Copper Flask, Lined with Silver, to show spheroidal state of water	0	15	0
Marcet's Steam Boiler, complete	5	5	0
Flask, with stopcock, to show ebullition of water under diminished pressure	0	5	6
Geissler's Patent Vapormeter, for ascertaining the quantity of alcohol in wine, &c.			
Brequet's Metallic Thermometer	4	0	0
Thermometer, in glass tube containing water, to show development of heat on its freezing	0	16	0
Syringe, with glass barrel, for igniting gases, &c. . . .	1	0	0
Candle Bombs . . . . per doz.	0	0	6
Rupert's Drops . . . . ,,	0	2	0
Bolognean Flasks . . . . ,,	0	6	0
Hero's Rotary Engine.			
Wollaston's Apparatus to illustrate the ordinary condensing engine . . . . 7s. 6d. and	0	10	6

	£	s.	d.
Tin Vessel and stopcock, to illustrate the condensation of steam and pressure of the atmosphere	0	5	6
Oscillating Engine, Working Model . . . . from	1	15	0
Working Models of Steam Engines made to order.			
Sectional Models of Steam Engines . . . . from	5	0	0
Zinc Ethyl Fountain Apparatus	0	18	0
Apparatus to show the compressibility of Liquids and liquefaction of Gases under pressure	10	10	0

PHOTOGRAPHIC CAMERAS & APPARATUS.

MODELS OF INVENTIONS,
AND ALL KINDS OF APPARATUS, MADE TO ORDER.

Wholesale and Shipping Orders executed with despatch.

Orders from Foreign parts must be accompanied by a Remittance, or Order for payment in London.

Post Office Orders to be made payable in Regent Street, W.

The greatest care will be taken in the packing of Goods, to prevent breakage, but W. L. will not hold himself responsible for damage done during transit.

Packing Cases charged Cost Price, and NOT *allowed for if returned.*